T0281196

Cambridge Elements ≡

Elements in the Philosophy of Biology
edited by
Grant Ramsey
KU Leuven
Michael Ruse
Florida State University

HOW TO STUDY ANIMAL MINDS

Kristin Andrews
York University

CAMBRIDGE
UNIVERSITY PRESS

CAMBRIDGE
UNIVERSITY PRESS

University Printing House, Cambridge CB2 8BS, United Kingdom

One Liberty Plaza, 20th Floor, New York, NY 10006, USA

477 Williamstown Road, Port Melbourne, VIC 3207, Australia

314–321, 3rd Floor, Plot 3, Splendor Forum, Jasola District Centre, New Delhi – 110025, India

79 Anson Road, #06–04/06, Singapore 079906

Cambridge University Press is part of the University of Cambridge.

It furthers the University's mission by disseminating knowledge in the pursuit of education, learning, and research at the highest international levels of excellence.

www.cambridge.org
Information on this title: www.cambridge.org/9781108727464
DOI: 10.1017/9781108616522

First published 2020

A catalogue record for this publication is available from the British Library.

ISBN 978-1-108-72746-4 Paperback
ISSN 2515-1126 (online)
ISSN 2515-1118 (print)

How to Study Animal Minds

Elements in the Philosophy of Biology

DOI: 10.1017/9781108616522
First published online: June 2020

Kristin Andrews
York University

Abstract: This Element examines the methods of comparative psychology, which remains especially concerned about how to study animal minds and behavior without falling prey to fuzzy thinking. Training in comparative psychology places special emphasis on avoiding bias and on avoiding developing warm relationships with animal subjects. The principles of comparative psychology, including anti-anthropomorphism, Morgan's Canon, rules to avoid forming relationships with animals, and the instruction not to presume anything about animal consciousness, have been introduced to minimize bias in the science. Rather than seeing animals as sentient beings who live in community and have their own interests, these principles instruct scientists to remain distant and detached, introducing different biases that students are not instructed to watch out for. In this Element I argue that rather than attempting to avoid bias, comparative psychologists should acknowledge a range of biases and seek widespread collaborative projects to integrate different approaches to studying animal minds.

Keywords: animal minds, comparative cognition, consciousness, comparative psychology, bias, apes

ISBNs: 9781108727464 (PB), 9781108616522 (OC)
ISSNs: 2515-1126 (online), 2515-1118 (print)

Contents

Introduction

The birth of a new science is long, drawn out, and often fairly messy. Comparative psychology has its roots in Darwin's *Descent of Man,* was fertilized in academic psychology departments, and has branched across the universities into departments of biology, anthropology, primatology, zoology, and philosophy. Both the insights and the failings of comparative psychology are making their way into contemporary discussions of artificial intelligence and machine learning (Chollett 2019; Lapuschkin et al. 2019; Watson 2019). It is the right time to turn a philosophical lens onto the methodologies of comparative psychology. That is the aim of this Element.

Comparative psychology is the umbrella covering the different ways scientists study animal mind and behavior. Comparative psychologists study animal behavior and mentality, including the mechanisms and inner states that allow crows to form hooks, vervet monkeys to give warnings, crabs to make trade-off decisions, and humans to use language. While focus is on the similarities and differences between different species, capacities are often studied in terms of their evolutionary history, development, and current ecological or cultural context. What this entails is that comparative psychologists have a range of different kinds of training, different areas of expertise, and different research questions. One comparative cognition researcher compares children and dogs on causal reasoning abilities (Daphna Buchsbaum), another looks at memory in corvids (Nicola Clayton). An animal behaviorist examines raccoon territories in a city (Suzanne MacDonald). One anthropologist looks at war in wild chimpanzees (John Mitani), another looks at how innovations are adopted in capuchin monkeys (Susan Perry). One ethologist looks at play in wolves (Marc Bekoff), and another looks at deception in birds (Carolyn Ristau). One biologist looks at male alliances in dolphins (Richard Connor), and another examines economic game performance in apes and monkeys (Sarah Brosnan). An animal welfare scientist studies the effect of enrichment on mink (Georgia Mason). A learning theorist looks at gambling in pigeons (Thomas Zentall). A zoologist studies social learning in bees (Lars Chittka). It is not easy to guess in which department you'll find a comparative psychologist.

This Element aims to examine the methods of comparative psychology, which remains especially concerned about how to study animal minds and behavior without falling prey to fuzzy thinking. Animals can be cute, so humans are often intrinsically drawn to them; and this love of animals is sometimes taken to be at odds with being a careful scientist. The methods of comparative psychology reflect this worry with their special emphases on avoiding bias and on avoiding developing warm relationships with animal subjects. The first two sections analyze three textbooks to examine what young scientists are taught. Section 1 investigates three methodological principles students are taught to follow: Anti-anthropomorphism, Morgan's Canon, and Anti-anthropocentrism. I argue that the first two principles should be discarded, and that developing relationships with animals should no longer be discouraged, as relationships promote understanding. Building on this critique, in Section 2 I challenge the common prohibition against including animal consciousness in comparative psychology by arguing that it does not harm the science, but promotes it. Setting aside the special prohibitions for comparative psychology, in Section 3 I examine how the quest for objectivity in comparative psychology introduces its own bias and argue that the different disciplines of comparative psychology will introduce different biases; the best scientists can do is identify the sources of bias. The quest to eliminate all bias is a misguided one. Section 4 applies these issues to recent debates in ape cognition research between scientists who work in the field and those who work in the lab, and suggests best practices for integrating knowledge from both sources.

I have had the opportunity to work with a number of comparative psychologists in a variety of contexts, including a stint at Lou Herman's dolphin communication lab in the 1990s, collaborative studies with Peter Verbeek on the child's theory of mind at the University of Minnesota's Institute for Child Development, conducting research on orangutan pantomime communication with Anne Russon at her field sites in Borneo, research on rat social learning with Noam Miller at Wilfrid Laurier University, and co-teaching a field course in dolphin communication with Kathleen Dudzinski. As a philosopher of science who has been in the field and in the lab, I have a perspective from which to compare the comparative psychologists. This Element is my attempt to synthesize almost thirty years of experience and thinking about how to study animal minds.

1 Methods of Comparative Psychology

1.1 Looking at Textbooks

There may be no better way to understand a science than to read the field's textbooks (Giere 1988), since textbooks are where a mature science typically states its general principles, theories, and methods (Kuhn 1962). I asked the comparative psychologists I know what they teach when introducing students to the field, and most of them told me they rely primarily on journal articles. The books that were mentioned included a few anthologies, monographs, and three textbooks: Sara Shettleworth's *Evolution, Cognition, and Behavior* (2010b, 2012), Clive Wynne's *Animal Cognition: Evolution, Behavior and Cognition*

(2004, and Wynne and Udell 2013), and John Pearce's *Animal Learning and Cognition: An Introduction* (2008). While the lack of reliance on textbooks suggests that there may be no overarching theory of comparative psychology, a review of these books shows that there is a unified methodology. The methods share the intended function of defending against bias and fuzzy thinking in the study of animal minds and behavior.

The textbooks teach students three methodological principles – Anti-anthropomorphism, Morgan's Canon, and Anti-anthropocentrism.

Anti-anthropomorphism – Rejecting "the attribution of human qualities to other animals, usually with the implication it is done without sound justification." (Shettleworth 2010a, 477)

Morgan's Canon – "[I]n no case is an animal activity to be interpreted in terms of higher psychological processes, if it can be fairly interpreted in terms of processes which stand lower in the scale of psychological evolution and development." (Morgan 1903, 292)

Anti-anthropocentrism – Rejecting "[holding] the human mind [to be] the gold standard against which other minds must be judged" (Povinelli 2004, 29), or "the incorrect and misleading notion of a phylogenetic scale or *scala naturae*." (Shettleworth 2010b, 18)

The first two of these principles have been subject to much scrutiny and criticism from philosophers of science, which should worry comparative psychologists. I will argue that the first two principles should be discarded and replaced with standard scientific methods. There isn't anything special about comparative psychology that requires these principles. However, Anti-anthropocentrism remains a useful principle, as it instructs us to remember that animals are not little humans dressed in furry, scaled, or feathered suits. Animals all have their own species-specific perspectives, goals, interests, and practices. Like the early anthropologists and ethnologists who had to learn new cultured ways of seeing people, comparative psychologists have to learn contraspecific ways of seeing animals.

1.2 Anti-anthropomorphism

Students of comparative psychology are taught that anthropomorphism is a bias that we have to overcome and has no place in science. We find prohibitions against anthropomorphism in all three textbooks.

In his critique, Clive Wynne writes, "[A]nthropomorphism is not a well-developed scientific system. On the contrary, its hypotheses are generally nothing more than informal folk psychology, and may be of no more use to the scientific psychologist than folk physics to a trained physicist" (Wynne

2004, 606). He claims that progress in animal mind science will benefit from having "explanatory frameworks that are concrete and unambiguous" and that anthropomorphism cannot offer anything close to that.

John Pearce instructs students on the perils of anthropomorphism:

> The temptation to attribute human feelings and experiences to animals remains to this day. Itani (2004, p. 228) provides the following quote from a psychologist called Roger Fouts, who works with chimpanzees and who communicates with them using sign language . . . The most famous of these chimpanzees is Washoe. Fouts (1997) states: "When I looked into Washoe's eyes she caught my gaze and regarded me thoughtfully, just like my own son did. There was a person inside that ape 'costume.' And in those moments of steady eye contact I knew that Washoe was a child." To labor the point that has just been made, it is possible that Washoe has similar mental experiences to a child, but it is also possible that Washoe has a very different type of mental experience, or no mental experience at all. Gazing into her eyes will not resolve this issue. (Pearce 2008, 24)

Sara Shettleworth writes, "Anthropomorphic interpretation of anecdotes like Romanes's story of the cat, that is, *folk psychology* (our everyday intuitive understanding of human psychology), suggests that animals should learn by copying others. But no matter how plausible the proposed explanation of a single observation or set of observations, there are nearly always other equally plausible explanations" (Shettleworth 2012, 4).

We see here two ways of understanding the Anti-anthropomorphism principle. The more specific interpretation, which two of the authors explicitly endorse, is an instruction to avoid folk psychology. I will show that folk psychology plays an essential role in comparative psychology and is the starting point, but not the end point, of research. The less specific interpretation of the principle is, as Shettleworth puts it, to avoid unjustified attributions of psychological properties. While standard scientific method agrees that scientists should avoid unjustified claims, I think the principle does more harm than good, and it should be discarded in favor of more general principles of scientific investigation.

1.2.1 Anti-anthropomorphism as Avoid Folk Psychology

Two of the textbook authors identified anthropomorphism with folk psychology. This conflation is typical in the literature. In a report authored by many leading animal cognition researchers (and two philosophers[1]), the authors worry that "Folk psychology is the linguistic equivalent of giving guns to children and telling them to play carefully: misuse is inevitable" (Jensen et al. 2011, 274).

---

[1] I am one of the authors of this report, but I never endorsed this claim!

Derek Penn accuses folk psychology of "ruining" comparative cognition (Penn 2011) and thinks that the "insidious role that introspective intuitions and folk psychology play" in comparative cognition research must be eliminated (Penn and Povinelli 2007, 732). Though folk psychological explanations may be "simpler for us" to understand (Heyes 1998, 110), Heyes thinks that they cannot be a part of science, since they commit a double error: first, we produce an unscientific explanation of human behavior, and then we apply that explanation to nonhumans when they engage in superficially similar behavior.

This conflation of "folk psychology" and "anthropomorphism" leads to confusion about what is really of concern. Let's clarify what is meant by folk psychology in the philosophical literature. Paul Churchland introduced the term, defining it as the "commonsense conception of psychological phenomena" that consists of concepts including belief and desire, and he proposed we should eliminate such terms in favor of neuroscientific descriptions (Churchland 1981, 67). In contrast, functionalism in the philosophy of mind takes the concepts of folk psychology as constructs that have causal power within an interpretive framework (Lewis 1972), and intentional systems theory has folk psychological concepts and generalizations picking out real patterns of behavior but not physical causal elements (Dennett 1991).

Contrary to the claims of the psychologists, appeal to folk psychology in the functionalist or intentional systems sense is needed to make progress in comparative psychology. Every operationalized term introduced into animal cognition that I am aware of has its genealogy in a term of folk psychology. "Affiliative relationship" comes from "friendship"; "episodic-like memory" comes from "memory"; "aversive" comes from "fear" and "dislike." Folk psychological concepts allow us to categorize behaviors into types and investigate the causes of those types of behavior.

For example, in order to determine whether chimpanzees understand what others can and cannot see, we need to categorize different patterns of movement as behaviors of the same type. One movement might be a chimpanzee lifting a hand to hide his fear grin from a competitor, and another might be a chimpanzee seeking out only food that a dominant cannot see. These two movements share no formal properties, but to do science we can categorize them together as both being of the same type of behavior and then ask what a chimpanzee needs to understand in order to engage in these sorts of behaviors – in other words, does a chimpanzee have to know that others have a different point of view from themselves. If we cannot see functional similarities between movements of different geometries, we fail to consolidate behavioral types as objects of study.

The need to organize behaviors together for study is presented as an important part of the methodology of comparative psychology. Shettleworth suggests that to compare humans and other animals requires "looking for *functional similarity* of behaviors across species" (Shettleworth 2012, 7). She provides the example of how we can test recognition memory in a human by asking them if they saw an item earlier, and notes that we can test recognition memory in a mouse by training mice to choose the odor that was most recently encountered. Furthermore, Shettleworth goes on to claim that "the logic of functional similarity is a basic tool in comparative cognition research (Heyes, 2008), even though, as in the case of the honeybees, deciding what constitutes functional similarity between observable animal behavior and evidence of some interesting human cognitive process is not always straightforward or uncontroversial" (Shettleworth 2012). If finding functional similarities is a key methodology for comparative psychology, then it is necessary to permit scientists to propose and define functional categories, including categories in terms of psychological properties that some believe or suspect may be uniquely human. Scientists must be able to ask the right kinds of questions.

Following the advice to avoid folk psychology in comparative psychology would lead scientists to avoid all the mental concepts associated with our lay (sometimes true and sometimes false) understanding of the causes of human behavior. Scientific psychology begins with folk psychology, just as scientific physics is based on folk physics when it comes to things like speed and momentum, or the distinction between solids, liquids, and gases. With investigation, we can refine our concepts, or discard them if we find that they don't have a role to play.

Giving up folk psychology would turn comparative psychologists into eliminative materialists, making it impossible to compare humans and other animals without likewise adopting eliminativism regarding human psychology. Since psychological concepts have been instrumental in psychology, giving them up would be a true loss. Some explanations for behavior would be left unexamined, derided as unscientific and insidious. Comparative psychology students should not be taught to avoid making some comparisons, which would be the result of a strict following of the principle understood as *avoid folk psychology.*

1.2.2 Anti-anthropomorphism as Avoid Unjustified Attributions

Another way to understand this principle is as a direction to avoid unjustified attributions. As defined by Shettleworth, anthropomorphism is understood simply as "the attribution of human qualities to other animals, usually with the implication it is done without sound justification." On this presentation, the advice to avoid anthropomorphism doesn't offer any special guidance to the student of comparative psychology. Since scientists should always avoid unjustified attributions and explanations, the principle of Anti-anthropomorphism doesn't add anything informative about how to go about doing good science.

The main problem that has been raised with the Anti-anthropomorphism principle is that there is no preempirical way to identify anthropomorphic properties. If it is a property unique to humans, then we could agree that that property ought not to be used to explain nonhuman behavior. However, to justify the claim that a property is uniquely human is to say we looked but do not see it elsewhere. Such claims should be the result of scientific investigation, not principles for starting an investigation. Anti-anthropomorphism either begs the question or redundantly instructs us to avoid false attributions (Keeley 2004; Andrews 2015).

One may object that Anti-anthropomorphism does offer guidance that is useful to the student of animal psychology. Anti-anthropomorphism directs students *to operationalize vocabulary* – to avoid unscientific concepts, and *to not form relationships with subjects* – to avoid attributing unwarranted mental states and becoming too tender-hearted. Let's examine each instruction.

1.2.2.1 Operationalize Vocabulary. One piece of advice that is often followed is to define terms carefully, use scare quotes, or add "-like" when using mental or other folk psychological terms. New terms are created, such as "affiliative relations" rather than "friend" or "forced copulation" rather than "rape."

For example, in their investigation of episodic memory in animals, Clayton and Dickenson's 1998 paper on memory in scrub jays, which was published in *Nature*, consistently used the phrase "episodic-like memory" because the authors could show only the behavioral elements of episodic memory, not the hidden element, namely conscious experience. Though we cannot directly observe human conscious experience of memory, there is no comparable worry about taking the behavioral elements to be sufficient evidence of unqualified episodic memory in humans.

It is good to be careful with words and to define our terms. But if we invent new words for nonhuman animals and keep old words for human beings, then we are going to introduce unnecessary problems when trying to draw comparisons between humans and other animals.

As discussed above, if we reject the use of folk psychological terms, we are not relying on functional similarity and will run into difficulty categorizing human and nonhuman behavior together into a single category. By shearing off part of a concept, like the consciousness in a scrub jay's episodic memory, we are not using the same methods for both groups, rendering comparisons artificial. This can impact the progress of science. Consider Jonathon Crystal's research finding episodic memory in rats, which is being used to help find a treatment for Alzheimer's disease (Panoz-Brown et al. 2018). If the rats don't have true episodic memory in which they experience their memories, but merely have episodic-like memory, then rats would fail to serve as good models for what is most debilitating in Alzheimer's.

The problem is also evident in plant behavior research. Michael Pollan reports on the sixth annual meeting of the Society for Plant Signaling and Behavior (formerly The Society for Plant Neurobiology) in 2013 in which animal ecologist Monica Gagliano presented a paper called "Animal-like learning in *Mimosa pudica*." *Mimosa pudica* is also known as the sensitive plant; it folds its leaves when disturbed. Gagliano showed that after repeatedly being dropped, a *Mimosa pudica* plant habituates to the disturbance and stops folding its leaves (Gagliano et al. 2014). It isn't that the behavior is exhausted, because other sorts of stimuli, such as rain or ruffling, will result in the plant folding its leaves. At the conference Gagliano presented this finding as an example of habituation learning, suggesting that the plants remember that being dropped isn't worth the bother of responding.

Pollan reports that Gagliano's use of words like "learning" and "memory" were largely derided by conference attendees:

> On my way out of the lecture hall, I bumped into Fred Sack, a prominent botanist at the University of British Columbia. I asked him what he thought of Gagliano's presentation. "Bullshit," he replied. He explained that the word "learning" implied a brain and should be reserved for animals: "Animals can exhibit learning, but plants evolve adaptations." He was making a distinction between behavioral changes that occur within the lifetime of an organism and those which arise across generations. At lunch, I sat with a Russian scientist, who was equally dismissive. "It's not learning," he said. "So there's nothing to discuss." (Pollan 2013)

Pollan also reported that another scientist "suggested that the words 'habituation' or 'desensitization' would be more appropriate than 'learning'" (Pollan 2013).

Pollan also reported that Gagliano told him that her paper at that point had been rejected by ten journals, but not because of a lack of scientific rigor. Rather, the editors rejected her use of the term "learning," and she had refused to revise the language. This rejection occurred despite the fact that habituation learning is typically taught as a variety of learning in psychology classes. Pollan quotes Gagliano, "Unless we use the same language to describe the same behavior we can't compare it." This is as true of plants and animals as it is of human and nonhuman animals.

Scientists should use clear language and operationalize their terms when necessary. But for comparative psychology to flourish, the terms should not be initially defined so as to prohibit cross-species comparisons (such as requiring a brain for learning) and should not require a different level of warrant for applying the term to one species over another. As we will see, comparative psychologists have a special worry regarding animal consciousness, which I think they must set aside. The Anti-anthropomorphism principle may lead to such errors rather than protecting against them.

1.2.2.2 Avoid Forming Relationships with Subjects. Another direction that stems from Anti-anthropomorphism is to avoid forming relationships with animals, fearing that relationships will lead scientists to unscientifically interpret animal behavior. Whether the subject is a rat or a chimpanzee, scientists are taught not to get too attached to their research subjects. Lab animals such as rats who are bred to be scientific research subjects are often sacrificed at the end of the experiment, which might make forming relationships more difficult. In the field or in zoos, scientists are also sometimes warned to keep an emotional distance. When Jane Goodall violated this principle by naming the chimpanzees she observed at Gombe and by using gender pronouns instead of the more common "it" to refer to chimpanzees, she was roundly criticized. Scientists worried aloud that Goodall wasn't being objective (Midgley 2001).

In contrast, scientists who study human children spend time building a relationship with the subject, even if it is only a few minutes of playing. During my PhD research I spent a month at a lab day care before I was allowed to test the children. Building relationships was of paramount importance.

Failing to form relationships with social animals, that is, animals who are inclined to form relationships, means failing to understand or communicate with the animal. Without the right kind of social environment, the scaffolding and motivation that are part of an animal's psychological system may not be present. For example, I've argued that the claim that only humans engage in overimitation (e.g., Clay and Tennie 2018), or the copying of causally irrelevant actions to achieve a goal, is based on studies that don't take into account the relationship between observer and subject (Andrews 2020). Given the theory that overimitation functions to help children learn cultural norms, and the finding that children only selectively overimitate – for example, they overimitate same language speakers but not foreign language speakers (Buttelmann et al. 2013) – a good test of overimitation in animals will have to consider the relationships between the demonstrator and the subject. Recent studies find that dogs overimitate their caregivers (Huber et al. 2018) but fail to overimitate unknown researchers (Huber et al. forthcoming).

The result of neglecting attention to relationships is that the scientist fails to see the full functioning of the animal. Avoiding forming relationships with animals results in treating the animal as a subject rather than a participant in the research. Japanese primatologist Tetsuro Matsuzawa goes so far as to describe Chimpanzee Ai as his "research partner." In his lab scientists adopt a methodological approach called "participant observation" according to which there is a triadic bond between chimpanzee mothers who raise their own offspring and human researchers. His colleagues are encouraged to form relationships with their ape research partners (Matsuzawa 2007). Matsuzawa and his team are able to report positive results on chimpanzee cooperation and imitation, while other researchers who don't take a partnership approach to the research have reported null findings. It may come as no surprise that when I asked Matsuzawa if Ai ever overimitated him, he said she did. In fact, years before overimitation in apes became a topic of interest, Matsuzawa and his colleague reported that their chimpanzees would copy his irrelevant tool use (Myowa-Yamakoshi and Matsuzawa 2000).

For chimpanzees, relationships matter. Relationships may matter for lots of other species, too.

If scientists are building relationships with children in order to get them to perform, and we want to compare a child and a chimpanzee, scientists need to build these sorts of relationships with animals to get them to perform as well. This might mean tickling rats when handling them to build a positive association with that handler, calling in a favorite trainer when testing a dolphin, or developing long-term relationships with zoo-living chimpanzees. It isn't just mammals that human scientists can develop relationships with. Morgan Skinner, a PhD student who is working with garter snakes, told me, "I have had a favorite snake every year. My current favorite is all attitude until I pick her up and then she is super gentle . . . Whether it is real or not, I definitely attribute traits to certain snakes. I can tell you that some of them adapt to my presence much better than others. Some of the garter snakes won't eat with me around, whereas others recognize me as the 'food bringer.' Last year, my favorite was 'Fancy Diva.' Whereas other snakes would bolt at the first chance they got, Fancy Diva associated me with food. She would often wait for me at the front of the terrarium and would not try to escape when I opened it. Instead, she would just wait for me to feed her" (personal communication). Skinner's attitude toward snakes is reflected in his choice of research topics – snake social behavior.

Anti-anthropomorphism, insofar as it directs researchers to avoid relationships, should be rejected. In Section 2 I will argue that researchers should take animals to be conscious beings with their own interests, and as potential social partners. Thinking it inappropriate to have any kind of a relationship with a research subject ignores an important truth, that animals are sentient beings who live in community – and this is even true of so-called solitary animals.

Anti-anthropomorphism, insofar as it directs researchers to operationalize their terms, overreaches by requiring human concepts to be operationalized differently from concepts applied to animals. To truly do *comparative* psychology, the concepts must remain the same when investigated across species.

1.3 Morgan's Canon

The oft-quoted rule "in no case is an animal activity to be interpreted in terms of higher psychological processes, if it can be fairly interpreted in terms of processes which stand lower in the scale of psychological evolution and development" is Morgan's Canon (Morgan 1903, 292). Morgan's Canon is a methodological principle that requires unpacking.

Current textbooks present Morgan's Canon as directing us to prefer associative over cognitive explanations: "In contemporary practice 'lower' usually means associative learning, that is, classical and instrumental conditioning or untrained species-specific responses. 'Higher' is reasoning, planning, insight, in short any cognitive process other than associative learning" (Shettleworth 2010b, 17–18). Of course, as Morgan wrote before the rise of behaviorism, *he* wasn't referring to associative learning principles.

Putting the author's intent aside, the first worry about Morgan's Canon is that no satisfactory account of what "higher" and "lower" mean has been offered (see, e.g., Allen-Hermanson 2005; Fitzpatrick 2008, 2009; Sober 1998, 2005; de Waal 1999). A related worry is that even if we had such an account, if we take Morgan's Canon to direct us, *ceteris paribus*, to accept a lower or associative explanation over a higher or cognitive explanation rather than remain neutral between the two explanations, we introduce a bias into the science (Fitzpatrick 2017; de Waal 1999; Sober 2005; Andrews and Huss 2014). Typically, when there is not enough evidence to decide between two explanations, we should remain open-minded until enough evidence comes forth. Simplicity may be a virtue of a scientific theory, but in the case of Morgan's Canon we lack a compelling argument for why some of these explanations would be simpler than others. One person's simplicity is another person's complexity. For example, Andrew Whiten argues it is simpler to explain chimpanzee mind-reading behavior in terms of mentalistic intervening variables that provide unifying links between the chimpanzee's beliefs about the world and the behavior of other chimpanzees. Mind-reading in terms of mentalistic intervening variables, he thinks, more simply explains chimpanzees' ability to categorize distinct behaviors as being of the same sort (Whiten 1994, 1996, 2013). On the other hand, Cecilia Heyes points out that while mentalistic attributions may be "simpler for us" (2008, 110), they are not simpler in general. In his analysis of this debate, Elliott Sober suggests that the data does not currently support a simplicity argument for either side (Sober 2015).

A third worry with taking Morgan's Canon to direct us to prefer associative over cognitive explanations is that since the rise of connectionist architectures and McLaren and Mackintosh (2000)'s theory of selective learning and stimulus preexposure, some forms of associative learning appear to be best understood in terms of representations, such that stimulus representations are what enter into associations. If there is no association/representational distinction, then we cannot understand Morgan's Canon as telling us to prefer associative explanations over representational ones. What about preferring associative explanations over "higher" explanations in terms of reasoning, planning, or insight? Here again, there appears to be no clear distinction, with even insight sometimes analyzed in terms of associative processes (Shettleworth 2010a).

The widespread existence of associative learning across species should not by itself justify the claim that there is not a cognitive explanation for the behavior. Mike Dacey argues that associative models do not describe some particular type of "associative" process but are abstracted partial descriptions of the causal relations that result in the behavior to be explained (Dacey 2017). This makes associative models compatible with cognitive processes, merely presented at different levels of abstraction. The choice of level depends on what we aim to do with the model. When we are comparing different species, a comparative psychologist will want to keep the level of abstraction the same.

A fourth concern is that by instructing scientists to prefer simplicity the Canon is undermining scientific progress. Irina Mikhalevich draws our attention to how Morgan's Canon directs psychologists to develop and test simple models, taking resources away from the development and testing of more complex models (Mikhalevich 2017). While it may look like the simpler models are more successful than the complex ones, she argues that this is due to a disproportionate allocation of resources into simple models. Mikhalevich describes how the creation of associative models that explain an animal's performance on tasks that had previously been interpreted in terms of mind reading, metacognition, or mental maps is taken to be evidence against those explanations. She shows how scientists have dropped their claims to animals having more complex cognitive capacities in light of a workable associative model, without first developing more complex models and comparing explanatory power. With simple models at hand, the science progresses by redesigning experiments to test simple models, thus resulting in what appears to be greater evidence in favor of them. Until we give as much intellectual attention to more complex models, we won't be able to justify the preference for a simple one.

Given these four concerns about Morgan's Canon, we may ask whether Morgan's Canon offers any methodological instruction to students of comparative psychology, or whether it should be done away with or taught only as a historical curiosity.

Among philosophers of science there has been a growing move toward leaving Morgan's Canon once and for all. Elliott Sober writes that rather than Morgan's Canon, "[t]he only prophylactic we need is empiricism" (Sober, 2005, 97). Like with the response to Anti-anthropomorphism, the idea is that so long as we are engaged in standard empirical practices, we don't need a special rule that tells us to avoid false claims. Avoiding false claims is the bread and butter of sciences generally, or so we hope. Simon Fitzpatrick suggests doing away with Morgan's Canon and replacing it with *evidentialism*: "in no case should we endorse an explanation of animal behaviour in terms of cognitive process X on the basis of the available evidence if that evidence gives us no reason to prefer it to an alternative explanation in terms of a different cognitive process Y – whether this be lower *or* higher on the 'psychical scale'" (Fitzpatrick 2008, 242).

Before we give up on Morgan's Canon, we should consider how it has been applied in comparative cognition. The Canon is often taught in the context of the Clever Hans episode. Clever Hans was a Russian trotting horse who appeared to do mathematical calculations, read German, and recognize musical notes by tapping his hoof the correct number of times. While early twentieth-century audiences were convinced that Clever Hans knew how to add, Oskar Pfungst found that Hans's owner was inadvertently signaling Hans when to start and when to stop tapping his hoof. Here the lower explanation is presented as Hans's sensitivity to the cue, and the higher explanation is that Hans can add and read German.

This story is taught to students as a cautionary tale, to demonstrate the importance of designing experiments that do not permit cuing. If the experimenter doesn't know the right answer or can't see what the subject is doing, then they can't cue the subject. For this reason, experiments with animals sometimes have the experimenter wear a welding mask or opaque goggles so that they can't see what the animal is doing, or ask a dog owner to sit perfectly still in the testing chamber while staring at a spot on the wall.

As we will see in Section 4, some scientists worry that the emphasis on avoiding cueing isn't held constant for human and nonhuman subjects. Furthermore, we can see some ways in which the responses to Clever Hans worries can adversely impact the science. In practice, avoiding cues can make for artificial testing environments because a human tester behaves oddly – wearing a helmet or not making eye contact – and yet the animal is supposed to perform at peak. Some behaviors require social scaffolds; a child may be more at ease to cooperate with a researcher if their parent is in the room or if they play with the researcher for a few minutes before the test. An animal may prefer to perform for a favorite trainer who alone can elicit behaviors given the quality of their relationship. Avoiding all possible cues can also take away scaffolds that facilitate the behavior. Subjects may be capable of a behavior given a richer context, but take the context away, and you impair the behavior. Consider your performance on an eye exam at the optometrist. It is harder to decode the letters because they are lacking any meaningful context; this is why getting a letter or two wrongish (e.g., saying O rather than C) doesn't result in a change in prescription.

Colleagues have told me that if they cannot figure out how to run an experiment without a feature that might serve as a cue, they often don't run it. In such cases we lose potential sources of information about whether the animal can perform with the cue in place. With that knowledge, we could continue testing and vary the cue to determine when it impacts the results. Without that kind of work, it would be difficult to tell the difference between a cue and a scaffold.

Furthermore, not all cues can be controlled for; while visual cues are the ones most salient for humans, other species may be attuned to different sensory modalities. If the protection against Clever Hans effects leads to doing fewer studies, then the advice should be very carefully considered before deciding that certain kinds of studies don't get to count as part of the practice of science.

Testing all relevant variables is advice relevant to any science looking at constructing causal models or uncovering the mechanisms supporting a phenomenon. Advising students to avoid introducing irrelevant variables, like advising students to avoid false explanations, is not advice that is unique to comparative psychology. However, what the Clever Hans case suggests is that there are types of variables that the student of comparative psychology should look out for. These variables are those that are based on the relationships subjects might have with the experimenter or other human who knows the correct answer. Hans had a relationship with his owner, relying on him to get the correct behavior. If we see Hans as trusting his owner, then attending to such relationships of trust is necessary for researchers studying animals, especially those dependent on adult humans.

This doesn't leave Morgan's Canon standing as a methodological principle of special interest to comparative psychology. Rather, the lesson of Clever Hans is to follow the third principle, Anti-anthropocentrism, and work to see the world from the eyes of the subject.

1.4 Anti-anthropocentrism

This final principle in comparative psychology has received less attention from philosophers of science, perhaps because it is less problematic. This principle is a prohibition against anthropocentrism: "[holding] the human mind [to be] the gold standard against which other minds must be judged" (Povinelli 2004, 29) or taking humans to be the "center of the universe" (Wynne 2004, 9).

In comparative psychology there is a universal explicit condemnation of anthropocentrism, which often comes with an endorsement of evolutionary theory. Wynne writes, "Darwin's achievement is to let us see that we are all machines, mankind included" (Wynne 2004, 9). Pearce's textbook doesn't include anthropocentrism in the index, though he is scathing in his presentation of the Great Chain of Being, or as he calls it, the "phyletic scale" way of thinking: "it is not difficult to regard the phyletic scale as roughly corresponding to the intellectual development of the species ordered along it. This interpretation could hardly be more incorrect" (Pearce 2008, 5).

Gordon Burghardt's defense of Anti-anthropocentrism reflects his endorsement of "critical anthropomorphism," which amounts to seeing the world from the animal's point of view. Burghardt is inspired by the biologist Jacob von Uexküll's concept of an animal's umwelt – the select features of their environment that animals relate to and engage with – and quotes him approvingly: "Our anthropocentric way of looking at things must retreat further and further, and the standpoint of the animal must be the only decisive one" (Burghardt 1991, 53).

The positive direction that comes from Anti-anthropocentrism is to attempt to see the world from the animal's point of view. Von Uexküll described the umwelt of the tick as the world as perceived by and relevant to a tick (von Uexküll 1957). The ethologist Nikolaas Tinbergen is said to have once described his work as interviewing the animal in its own language. Thomas Nagel wondered what it is like to be a bat (Nagel 1979). Dorothy Cheney and Robert Seyfarth titled their book on vervet monkey cognition *How Monkeys See the World* (1990). The Anti-anthropocentrism principle is reflected in these endeavors to recognize what is salient to the animal species being studied.

This advice is well taken when it comes to animals and humans alike. Cultural anthropologists recognize that different aspects of the environment will be salient to individuals of different cultural groups. When studying a culture, we must know what is relevant to members of that community. When we try to describe a culture by looking only at what is relevant to us, we end up missing important information that helps us to understand that culture, whether it is a human or nonhuman one. Anthropologists recognized this long ago (Goodenough 1956).

1.5 Conclusions

The three principles of Anti-anthropomorphism, Morgan's Canon, and Anti-anthropocentrism are part of the history of comparative psychology, but the first two should be discarded as outdated and unhelpful. While the intention behind these principles is to avoid bias and fuzzy thinking, adopting them substitutes one bias for another. Following these principles can create the bias of anthropectomy (Gk. *anthropos* – human; *ectomy* – to cut out) – denying human properties to animals when they may indeed exist (Andrews and Huss 2014). As we will see in Section 3, bias is part of science, and instead of trying to avoid it at all costs, we can seek to recognize it where it stands.

To do good science, scientists can use terminology consistently across species, be clear about what their terms mean, and recognize that their interactions with subjects can impact their findings. Rather than avoiding building relationships with their subjects, students should be instructed to do their best to gain the trust of the animals they work with. Students should also be taught to attend to the differences between species in terms of their ecological and social umwelts.

In comparative psychology, experiment and observation have a clear impact on the subject. Whether it is a pigeon in a lab or an orangutan hiding in the forest, human observers are part of the situation. There is no denying that there are relationships between scientists and their animal participants, even if they are all business. An attempt to take the human out of the picture is to set an impossible goal and to mislead about the objectivity of the results.

The relationships will vary dramatically, of course, but just as developmental psychologists treat children as participants in research, scientists should take the animals they work with as participants, not subjects of study that are immune to the care and attention of the humans around them. This way of thinking about the case of Clever Hans offers a deeper and more complex picture of how scientists and subjects are interrelated and reminds us that some capacities can only be expressed in certain kinds of contexts because some behavior is not merely internal to the organism but emerges from being embedded in a certain ecology or social structure.

What emerges from this discussion is the idea of taking seriously the fact that the research subjects in comparative psychology, as proper social partners who have their own perspectives on the world, are sentient beings. I can anticipate an objection to this advice – it leads to fuzzy thinking to take animals as social partners, since that requires seeing them as conscious beings. Consciousness, or sentience – the capacity to feel – has been largely left out of the textbooks as unscientific and harmful to the practice of science. In the next section I will argue that comparative psychology should accept animals as conscious beings. Trying to understand what animals see, feel, and think will facilitate research. Recognizing this can help us to ask rich questions about animal capacities and better get to know "the 'endless minds most beautiful' (Finlay 2007) of the other creatures on this planet, what they share and how they are unique, how and why they might have evolved" (Shettleworth 2012, 1).

2 Conscious Animals in Comparative Psychology

2.1 Conscious Mind

The defensive attitude that supports the principles of comparative psychology has also led to a general tendency to keep consciousness out of the investigation of animal minds. Phenomenal consciousness, sentience, or Nagel's "what it is like" is largely ignored in comparative psychology, both as a subject of study and a variable to consider. By "consciousness" I am referring to things like sensory experiences, imagery experiences, vivid emotions, and dreams. I am not referring to things like hormone release, dispositional knowledge, standing intentions, or responses to masked sensory stimuli (Schwitzgebel 2016). I am also not referring to cognitive capacities such as metacognition, planning, or theory of mind.

Consciousness in comparative psychology did enjoy some prominence in the late twentieth century. Donald Griffin's call to make the scientific study of animal consciousness respectable came in his 1976 book *The Question of Animal Awareness,* and it was taken up by a group of scientists who called themselves "cognitive ethologists." However, by the beginning of this century the movement was largely abandoned. As Colin Allen (2004) notes, very few scientists embraced the label "cognitive ethology."

When we look at the textbooks, we see that when consciousness is mentioned at all, it is presented as an inappropriate topic. This is a flaw. Ignoring consciousness harms comparative cognition. Along with Anti-anthropocentrism, students of comparative psychology should be taught how to consider animal consciousness.

Let's start by looking at what the textbooks say. In John Pearce's *Animal Learning and Cognition: An Introduction* there is only one entry for consciousness in the index. Pearce is critical of psychologist Robert Hampton's claim that rhesus monkeys might be conscious given their performance on a memory monitoring task, writing, "Hampton's evidence is certainly consistent with the claim that monkeys are conscious of some of their mental processes, but we are frustrated by the lack of methodology to determine if this is actually true because it is not possible to observe directly the mental states of an animal" (Pearce 2008, 221). Pearce suggests that animal cognition researchers may never have the scientific tools necessary for drawing conclusions about consciousness, writing, "Whether these tools will ever be adequate is a matter for debate, which – I suspect – will be waged for many years to come" (Pearce 2008, 169).

Pearce expresses a stronger skeptical view about animal consciousness in his defense of Anti-anthropomorphism; recall that he wrote, "it is possible that [chimpanzee] Washoe has similar mental experiences to a child, but it is also possible that Washoe has a very different type of mental experience, *or no mental experience at all*" (Pearce 2008, 24) (my italics). Here Pearce not only expresses concern with the methods for determining the qualities of an animal's experience, but he raises the question of whether animals feel anything at all. This position is not unique to Pearce; I have heard comparative psychologists say, "maybe the rat doesn't feel *anything*." Such statements suggest a deep skeptical worry along the lines of the problem of other minds.

The other textbook authors don't go so far as to doubt animal consciousness, but they do dissuade students from considering or investigating animal sentience. In Clive Wynne and Monique Udell's textbook *Animal Cognition: Evolution, Behavior and Cognition* there is also only one entry for consciousness in the index. Here they approvingly quote Daniel Dennett on how the consciousness debate is muddled in philosophy: "It seems positively foolhardy for an animal psychologist to blunder in where even philosophers fear to tread" (Wynne and Udell 2013). However, they go on to say that some psychologists, "having been appraised of the risks," are willing to confront the challenge of consciousness. Studying consciousness in animals is presented as a quest for heroes rather than the practice of normal science.

In her textbook *Cognition, Evolution, and Behavior*, Sara Shettleworth raises the concern that without the ability to tell us what they feel or want, we cannot be justified in describing how or what they feel:

> [B]ecause evidence for consciousness in humans generally consists of what people say about their mental experiences, seeking it in nonverbal species requires us to accept some piece of the animals' behavior as equivalent to a person's verbal report ... Therefore, the point of view of most researchers studying animal cognition is that how animals process information can, and should, be analyzed without making any assumptions about what their private

experiences are like ... This approach takes support from evidence that people act without ... using reflective consciousness, more often than is commonly realized. (Shettleworth 2010, 7)

The argument seems to go like this: Because language gives us evidence of the content of experience in humans, and other animals don't use language, we have no equivalent warrant for any claims about what animals feel. For this reason, scientists ought not concern themselves about what their experience might be like.

In all three of these textbooks, the authors raise concerns about a role for consciousness in comparative psychology. This is in contrast with recent philosophical debates about whether animals such as bees, crabs, fish, and even plants are conscious (e.g., Allen and Trestman 2017; Calvo 2016; Tye 2016).

In contrast, I think that consciousness should be an essential part of comparative psychology. Parallel to moral arguments that we ought to premise animal consciousness in making decisions about our treatment of animals (e.g., Birch 2017), I offer an epistemic argument that we ought to premise animal consciousness when studying animal minds and behavior. Rather than a Precautionary Principle for animal welfare, I defend a Curative Principle for comparative psychology. The Curative Principle can be stated as follows: When ignoring sentience hinders the ability to generate new knowledge of animal mind and behavior, and there is the potential to generate new knowledge by premising sentience, scientists ought to do so.

First, I offer five ways in which presuming animal consciousness can promote the science of comparative psychology:

a) Increase the number of topics open to study, including methods for studying the quality of conscious experience
b) Better recognition of relevant variables
c) Better creation of eliciting conditions
d) Better capacity to judge areas of continuity/discontinuity
e) Better welfare considerations

I then turn to the arguments against including consciousness in comparative psychology. The first argument I consider is the *skeptical argument*, according to which the classic problem of other minds is more an issue for animal minds than it is for human minds. I then present and critique two arguments found in the literature for why scientists need not consider animal consciousness. What these arguments share is an epistemic worry about the difficulty or impossibility of determining what animals feel. The *agnostic argument* suggests that we can fully account for animal welfare while remaining silent on animal consciousness. A central claim of this argument is that nothing is lost to the science by avoiding consciousness. The *language argument* defends the epistemic concern that while language gives us access to human conscious experience, there is nothing akin to language that will allow us to understand what other animals feel. I respond to this claim by pointing to a variety of methods that we can use to determine the content of an animal's experience.

With good reasons to accept animal consciousness into the science and no strong arguments to the contrary, comparative psychologists should be released from the prohibition to consider what animals feel. A presumption of sentience plays an essential role in much animal research, from the use of motivation and punishment in experiment to the study of affective states such as pain and emotions. Consciousness is also an implicit assumption in some current research programs, such as those concerning metacognition, emotion, and episodic memory. There is value in making this presumption explicit and training students how best to handle animal consciousness.

2.2 The Harm of Ignoring Consciousness

I can think of five areas in which one's commitment to consciousness can impact the practice of comparative psychology: topics of study, recognition of relevant variables, constructing eliciting conditions, evidential basis for making comparisons with humans, and welfare considerations.

Some topics of study cannot be investigated without granting that the subject feels something, and for other topics of study it is important to understand what an animal feels. If animal consciousness is deemed unscientific, then either these topics get excluded from investigation or they are only partially investigated, with the conscious elements being left out. Research on pain, emotion, pleasure, the phenomenal effects of analgesics and anti-anxiety drugs, and the cognitive biases of pessimism and optimism are all grounded in largely unstated assumptions about animal consciousness. Without permitting the premise that animals are conscious, we would have no reason to investigate grief, empathy, depression, or moral sentiments in animals. We can only investigate episodic-like memory rather than episodic memory, which includes an element of conscious awareness. Intentional actions, such as communication or goal formation, also become difficult to study directly if conscious experience is excluded. When we study this phenomenon in humans, we presume the human subjects are conscious. Again, we need to keep our concepts and operationalized terms stable across species if we are going to do comparative psychology. Furthermore, if we deny or remain agnostic about animal consciousness, the science of consciousness studies will also be hindered, since research on perception and the neural correlates of consciousness got off the ground with the assumption that monkeys are conscious perceivers (Crick and Koch 1990).

A related worry is that scientists who do not consider animal consciousness may overlook relevant variables and eliciting conditions in experimental contexts. Sensory experiences such as a bad dream, a depressive state, or fear of something in the environment may negatively impact the performance of a subject in an experiment. Using Anti-anthropocentrism, researchers who consider consciousness are free to imagine what variables might bother their research subjects and what variables might support them. Whether a subject is handled by a fearful experimenter or a calm one, or whether subjects are housed individually or in social groups are differences in methodology that may be relevant to a sentient being. I have worked both with scientists who expressed no concern about how individually housing rats would impact their performance on a social learning task, and with scientists who would invite the animal's favorite trainer to run a key experiment.

Recognizing a wider range of possible variables in an experiment can also help scientists design experiments that will better elicit the phenomenon they are investigating. For example, in the ape theory of mind research, scientists failed for almost forty years to elicit false belief tracking behavior. The stimuli and procedures used in these studies involved food hidden in boxes, and humans using a marker to indicate the location of the food (Call and Tomasello 1999). The failure of chimpanzees to pass the tests wasn't indicative of their abilities to track false beliefs, but illustrative of how thinking like a chimpanzee can help us better understand the chimpanzee mind. It is no surprise that the procedures that first elicited false belief tracking in apes came from the Primate Research Institute (PRI) in Kyoto and used stimuli showing someone dressed in an ape suit attacking an unsuspecting researcher (Kano et al. 2019; Krupenye et al. 2016). PRI is the home of primatologist Tetsuro Matsuzawa, who, as we saw, takes a participant-observation approach to research with apes and describes Chimpanzee Ai as his "research partner."

As was discussed in the previous section, the role of relationships between individuals has often been overlooked or only considered as a potential cue. When animals are not taken to be sentient subjects, the relationships between investigators and subjects are more easily overlooked. A scientist may not notice that their subjects enjoy working with some humans more than others, and they might not notice how the scientist is part of the situation or a relevant scaffold for the behavior. The subject's motivation to perform, or their quality of performance, may depend on their feeling toward the human who lifted them out of an enclosure or approached them at station. Furthermore, without considering animals to be sentient, the relationships between subjects may be more easily overlooked. In chimpanzees we have evidence that in studies involving dyads, such as social learning, imitation, cooperation, or inequity aversion tasks, the outcomes can be impacted given the relationship between the partners (e.g., Brosnan et al. 2005; Suchak et al. 2016). This may be true of other species as well.

A refusal to think about animal consciousness while at the same time unapologetically thinking about human consciousness will bias research both into the continuity of the mental across species and into attempts to find the kind of deep differences between humans and other animals needed to support human uniqueness claims. Human dominance is reinforced when we assume that humans are conscious, but remain agnostic about all other animals. Human practices that some take to be indicative of our special status on this planet, like insight, cooperation, culture, or morality, are going to be impossible to study from a comparative perspective when human conscious experience is part of the investigation and animal conscious experience is not.

A final consideration is that animal welfare may be compromised without a presumption of consciousness in animals, and we will have more to say about this when discussing the *agnostic argument*. Take the case of Harry Harlow's social isolation experiments, in which he separated infant monkeys from their mothers to discover treatments for human mental illness. Harlow used the monkeys as human models. In his autobiography, John Gluck, an animal ethicist who was trained by Harlow, describes Harlow's views about the monkeys' mental life:

> [Harlow] was a leader in demonstrating that rhesus monkeys were vulnerable to all kinds of harm, not just physical pain. He showed that monkeys could be emotionally destroyed when opportunities for maternal and peer attachment were withheld. He argued that affectionate relationships in monkeys were worthy of terms like *love*. In his work on learning in monkeys he vanquished the totally robotic view of the process offered by the behaviorists by offering abundant evidence that monkeys develop and evaluate hypotheses during attempts to develop a solution. Everything that Harlow learned from his research declared that monkeys are self-conscious, emotionally complex, intentional, and capable of levels of suffering. (Gluck 2016, 93–94)

The logic of the studies, which were explicitly focused on finding treatments for human depression and other mental illnesses, was based on the assumption that the monkeys could suffer the same kinds of mental anguish – conscious pain – as humans. When animal conscious experience is assumed, but not discussed, ethical concerns about the welfare and flourishing of these subjects are all too easily dismissed. The ethics of scientific practice may be compromised by a failure to presume animal consciousness.

Taken together, these five reasons to accept animal consciousness in comparative psychology should be compelling. Unless consciousness is epiphenomenal, that is, unless it has no causal power, a reluctance to investigate an entire domain of influence on animal behavior will lead to theories that are misguided, incomplete, or downright false. Furthermore, since consciousness already creeps into the science, instructing students how best to study animal consciousness will improve and systematize current practice. Unless there are good arguments against including consciousness in comparative cognition, these reasons should be taken as sufficient. Let's now turn to the three main arguments against.

2.3 The Skeptical Argument

The classic problem of other minds is a skeptical worry about other conscious minds based on the view that the only real evidence for conscious mind comes from one's own direct experience of their sentience. If scientists feel the pull of the skeptical worry when it comes to animals, they should realize that no empirical evidence gathered from a third-person perspective will ever be able to provide proof of a conscious mind, be it human or nonhuman. There are no tools that can bridge first-person experience and third-person observation. The skeptical problem will always hang there in the background. For this reason, the skeptical problem should not be seen as a problem for science. Science is grounded on background assumptions we cannot prove, such as the existence of an external world and the existence of other minds.

Scientists who study humans don't confront the skeptical worry. Psychologists and anthropologists are happy to assume that humans across cultures have rich inner lives. This is true even in communities where people do not talk about their mental states or sensations; those anthropologists who study human cultures that follow an "opacity doctrine" accept that it is difficult or impossible to know what is in others' "hearts and minds," but they do not deny or even question these people's conscious mental experience (Robbins and Rumsey 2008). From these fields we don't hear anything akin to Pearce's worry about Washoe, namely that it is possible a normally functioning human has no mental experience at all.

Because the skeptical problem about other conscious minds is not defeasible given empirical evidence, the problem falls outside the domain of science. Just as physics doesn't need to prove the existence of the external world and anthropology doesn't need to prove the existence of human consciousness, comparative psychology doesn't need to prove the existence of animal consciousness.

2.4 The Agnostic Argument

One argument denying a place for consciousness in comparative psychology comes from Marian Dawkins. Dawkins does not deny that animals likely have sentience, but she does think that the science of animal welfare benefits from avoiding talk of sentience. Since we currently lack an accepted account of consciousness, Dawkins worries that animal welfare science would suffer if it required reference to animal consciousness. She writes, "I want to stress that I am not denying consciousness in animals. For all we know, many species besides our own do have subjective feelings, possibly like ours, but possibly quite different. What I am arguing is that in the long run we will have a healthier biological approach to the study of consciousness if we acknowledge the uncomfortable, inconvenient and unsatisfactory truth that conscious awareness is still an unsolved problem" (Dawkins 2017, 3).

Dawkins argues that we can best protect an animal's welfare by not presuming that an animal is conscious, but by considering two objective measures of welfare: health and having what one wants (Dawkins 2008). Both of these measures are understood independently of consciousness. Health can be gauged by objective features such as longevity, a lack of pathological self-harming behaviors, and immune response. Having what one wants can be gauged through experimental tests, such as giving animals preference tasks and then providing them with what they choose, be it access to conspecifics or dark spaces in their enclosure.

We can consider two limitations with this approach. For one, Dawkins appears to be sneaking in the intrinsic good of conscious experience in order to defend the instrumental goods of health and having what one wants. From the Buddhists to the Stoics to Schopenhauer, health and having what one wants are both identified as morally relevant insofar as they protect us from suffering. Since suffering is a conscious sensation, Dawkins cannot use traditional arguments to promote health or having what one wants as a moral value.

Dawkins may not be concerned with this first problem, given that she isn't making a moral argument but rather a pragmatic argument about how best to promote animal welfare science. She can rely on the fact that humans already care about animal welfare in terms of health and having what one wants. Rather, Dawkins worries that appeal to animal sentience will impair progress toward the goal of animal welfare, because scientists who look for consciousness by measuring proxies such as hormones, thermal imaging, or activity levels are never measuring consciousness itself. Different scientists may interpret the same set of data as evidence that the animal feels calm or depressed, for example. She suggests that such practices would damage welfare science, since unwarranted interpretations would substitute for the hard data on health and having what one wants.

A second problem is that scientists may not be able to gauge an animal's health and what an animal wants without making some assumptions about what they feel. Let's take health first. Dawkins suggests we can judge health in terms of a long life and lack of self-harm. However, one may live a long life and not engage in self-harming behavior, but still be suffering. James Stockdale bore his seven years as a prisoner of war through Stoic practice, but in his writings he never described the experience as promoting his welfare. A long life is consistent with a life of pain.

Furthermore, there is a move in human health care to acknowledge the connection between mind and body. Such holistic approaches are not just found in alternative medicine; connections between mental state and physical health continue to be discovered in mainstream medicine. A depressed or stressed individual who shows no obvious signs of their current mental disorder is at risk for future health problems. For example, there is growing evidence that trauma and stress are correlated with the onset of autoimmune disease (Song et al. 2018). The relationship between mental well-being and a healthy body is probably why my family doctor asks me about my state of mind during each annual checkup. It is reasonable to think that knowing what promotes health for an animal will also require some thought about how they feel.

Likewise, knowing what an animal wants may not be possible without considering how an animal feels. Dawkins suggests we can come to know what an animal wants by giving them preference tasks, a choice between two options. However, I'm not sure how we can determine which options to provide the animal without having some prima facie consideration of what the animal wants. These assumptions about what animals feel are just starting places for doing preference tasks – they don't substitute for the research. Thinking about how animals might feel can help us get the science started.

For example, in a rat preference task scientists presume that a choice between a light and a dark enclosure is relevant, but a choice between television shows is not. This suggests we already know something about what an animal cares about. Any forced choice task, which a preference task is, presupposes something about the domain of plausible answers. To understand that an animal may want to be near a conspecific, that they will be motivated to work for high-value rewards, and even that they move away from aversive stimuli and toward attractive ones assumes something about the animal's mental experience. The

Anti-anthropocentric principle, which directs us to see the world from the point of view of the animal, is used in developing preference tasks. We don't test dogs to see if they prefer foods we like, but foods we think they might like. We don't test chickens to see if they prefer living as we do, but we consider how different roost heights may be of variable preference to a fledging chick and a mature chicken. To see the world from the perspective of the animal is to presume that they have a perspective – a point of view. This is already to presume that there is mental experience.

In addition, it is worth noting that the ability to make a choice does not entail that the choice promotes one's well-being or reduces suffering. A prisoner might have a choice between two different cellmates, but find them both oppressive. We often have to choose the least-worst option. Furthermore, since we don't always know what is best for us, having a choice may not lead to good welfare outcomes. A rat that is addicted to morphine might choose morphine over food, which doesn't promote their welfare. Having a choice is not the same as having good welfare.

I think Dawkins cannot avoid appeal to animal consciousness. Is there a way to include animal consciousness in comparative psychology without risking unwarranted interpretations of the data? I think so, and in what follows I show how good science and interpretation of data can go hand in hand.

2.5 The Language Argument

The language argument claims that we shouldn't study consciousness in animals because we lack a good evidential basis for what animals feel. Shettleworth suggests that the problem of studying consciousness is only a problem when it comes to other animals, not a problem when it comes to humans, because humans use language. Pearce makes a similar argument about the importance of language for determining mental states, though he expresses a special worry about animal honesty: "In the case of humans, we infer that someone is experiencing a state of uncertainty by asking them. But we are unable to ask monkeys about their mental states and, even if we *were* able to ask them, it would be impossible to know if they were telling the truth in their responses to our questions" (Pearce 2008, 169). I'm no more worried about lying monkeys than I am about lying humans. We are always interpreting when engaged in communication with others. If adopting the principle of charity and assuming others' communicative acts are meaningful permits robust prediction and coordination, then we have the best evidence we can have that the communicative acts mean what we think they mean – be they linguistic utterances or natural expressions.

The language argument suggests that linguistic behavior has a special status over other types of behavior as evidence for conscious experiences. The implication is that if we are to study consciousness in animals, we need to find an evidential base as powerful as language for other animals. But not all humans use language, and yet we still ask about, care about, and study their experience. Language is one evidential basis for consciousness, but there are others.

While the language argument is not a skeptical argument about the very existence of other minds, it has something in common with practical concern about how to determine whether some humans are conscious. For example, individuals with disorders of consciousness lack behavioral markers of consciousness, such as goal-directed movement or shrinking away from pinpricks, but still demonstrate neural function in areas of sensory processing, language processing, learning, pain responses, emotional responses, and recognizing familiar voices (see Johnson and Lazaridis 2018). The neural properties serve as evidence of consciousness to the medical professionals working with such individuals and are helpful in determining the types of treatments that might benefit them. The properties also help us to continue enriching our views about what counts as markers of particular conscious states.

Human infants also lack language, and during much of the twentieth century the medical community operated on infants without analgesics, given a widespread view that infants do not feel pain. It wasn't until 1987 that American medical professional organizations such as American Academy of Pediatrics decided to accept infant pain, with an editorial in the New England Journal of Medicine calling the evidence "so overwhelming that physicians can no longer act as if all infants were indifferent to pain" (Boffey 1987). Today there exist several diagnostic scales to measure infant pain, such as The Children and Infants Postoperative Pain Scale, which looks at pain markers such as facial expression, crying, restlessness, and bodily posture.

If we can determine what infants and humans with disorders of consciousness feel in some cases, then we have, in principle, methods to make justified claims about animal experiences. To defend the claim that comparative psychology should not involve itself with animal consciousness requires showing that there is a relevant difference in kind between the sciences of psychology or anthropology when they are focused on human subjects and when they are focused on animal subjects. Language cannot be the difference, because while language is not present in humans with disorders of consciousness and prelinguistic infants, scientists have managed to study and treat humans who don't use language.

The logic of the language argument fails with the introduction of other evidential bases for what animals experience. Language may be sufficient evidence, but it is not necessary.

Another approach is to study markers indicating types of consciousness, such as pain – the affective emotion associated with cellular damage (Allen et al. 2005; Carruthers 2004; Tye 2016; Shriver 2018; Varner 2012).

Pain markers that have been studied in animals include:

- Nociceptors connected to the brain or central nervous system
- Endogenous opioid releasing system
- Responsiveness to analgesics
- Self-administration of analgesics
- Nonreflexive behaviors such as nursing, favoring, or rubbing the damaged area
- Trade-offs to avoid noxious stimuli
- Responsiveness to noxious stimuli as punishment in learning tasks
- Direct stimulation of animals' brains successfully used as punishment

All mammals, birds, and reptiles that have been studied appear to have these markers, leading many scientists and philosophers to conclude that they have pain experience (e.g., Proctor et al. 2013; Tye 2017; Varner 2012). Other taxa, including fish, cephalopods, sea slugs, crustaceans, nematodes, leeches, and flies, have been studied and found to have some of these markers (see Sneddon 2015 for a review).

A marker approach could also be used in the study of animal pleasure. Pleasure is a response to reward that has positive affect. While animal pleasure is not as studied as animal pain, we can create the following list of markers that could serve as evidence of pleasure in animals:

- Neural pleasure pathways
- Dopamine production
- Responsiveness to dopamine interventions
- Self-administration of dopamine or direct stimulation of brain
- Nonreflexive behavior such as laughing or playing
- Trade-offs to approach rewarded stimuli
- Responsiveness to attractive stimuli as reward in learning tasks
- Direct stimulation of animals' brains successfully used as reward

Most of the research on pleasure has been focused on humans and rats, and we have evidence of many of these markers in rats, which have been subject to research on pleasure since the 1950s, when electrodes were implanted in the septal region of their brains. When the rats were given the opportunity to push a lever generating a signal to their "pleasure center," that's all they would do. Rats would trade off other important activities, including eating, drinking, sex, and nursing newborn pups, in order to stimulate their pleasure center (Olds and Milner 1954; Olds 1956). The rats would have starved to death had they not been disconnected from the apparatus.

Rats bred for the laboratory enjoy playing and will solicit play behavior with humans through "play bites." The longer the rats have been left alone in a cage, the more they seek play. Furthermore, rats laugh in ultrasonic chirps when they are tickled and when they play. After being tickled, rats bonded with the tickler and approached the tickler more frequently for social interaction. Rats who have been tickled are more cooperative research partners, and they prefer to spend time with other rats who laugh (Panksepp and Burgdorf 2003).

The examples of pain and pleasure show how we can gain knowledge of conscious states in other animals without relying on linguistic reports. Instead, using a marker approach and making judgments of conscious states given an overall body of evidence can serve as better evidence than mere linguistic behavior. The best indicators of consciousness will come in clumps, not singly. After all, we will rightly doubt an android who says "Stop, it hurts!" but fails to demonstrate any other behavioral marker.

2.6 Premising Conscious Mind

Once we reject the arguments for denying animal consciousness as part of normal comparative cognition, and see the benefits of including it, we can make better progress in the sciences of animals as well as the sciences and philosophy of consciousness. The Curative Principle's directive – when ignoring sentience hinders the ability to generate new knowledge of animal mind and behavior, and there is the potential to generate new knowledge by premising sentience, scientists ought to do so – suggests that the next generation of comparative psychology textbooks should pay sustained attention to issues of animal consciousness.

Without institutional support, consciousness research can't get off the ground, as Jaak Panksepp reports. Panksepp, who is credited with founding the field of affective neuroscience (Walker 2017), described how in the 1970s he was unable to publish his research on attachment in dogs, with reviewers rejecting his studies as "crazy" because he ascribed emotions to dogs. Panksepp said, "We must have written at least half a dozen grant proposals, and the message was clear: We're not gonna get funded no matter what we do. Dogs were the perfect species for the study of social attachment, but no one got it. The best canine behavioral research laboratory, and the last one in the country, died with me. I was incredibly disappointed" (Weintraub 2012). Today things are different. Just as infant researchers finally decided that infants do feel pain, and that heart surgery without analgesic is morally repugnant, dog researchers today are able to investigate dog emotion, social attachment, and sensation.

With the introduction of animal consciousness into comparative cognition, scientists can study the ways in which animals are conscious and construct experiments that are sensitive to how the procedure and methodology might impact their subjects. By training young scientists how to study conscious states in animals, we can protect against worries that such research is unscientific. Just as scientists have developed methods for studying and identifying the conscious states of infants and humans with disorders of consciousness, comparative psychologists can develop good methods for examining animals' conscious states.

Scientists will also be better placed to protect animal interests and well-being by learning what causes pain and pleasure behaviors in different species. We can get better results out of our experiments by treating animals as participants in the research, rather than as mere subjects of study. We can ask more questions, but we will also have to control for more variables. We need to keep doing – and funding – the research.

Accepting Anti-anthropocentrism and the Curative Principle in comparative psychology provides a foundation from which to do the research. These principles direct students and practitioners to take their subjects, be they fish, rats, dogs, or chimpanzees, as sentient beings with points of view, and with whom they are in relationship. While the principles may open up new areas for bias, that shouldn't be a deterrent. As we will see in the next two sections, bias is a part of science, and rather than hoping to find bias-free ultimate truths, scientists should seek to uncover and present the particular biases that arise from the methods being used. Section 3 will look at the quest for objectivity and the sources of bias in comparative psychology.

3 Objectivity and Bias in Comparative Psychology

3.1 Beyond Romantics and Killjoys

Like the study of animal minds, the study of human minds is robustly multi-disciplinary. Descriptions of the same phenomenon, say an autocrat's rise to political power, can be described within different fields. The descriptions will look different from one another, as each field focuses on different aspects of the case, and at a hasty glance these explanations may appear to be in conflict. A social scientist who thinks of humans as *Homo economicus* might explain the rise of autocrats in terms of self-interested individuals who willingly take part in autocratic systems, because the alternative – an anarchic state of nature – is so much worse. Such individuals believe it better to live under a strongman who has a monopoly on violence and a selfish reason to keep his vassals relatively happy than to be at the mercy of roving bandits who readily kill and plunder. This makes the choice a rational one. A biologist, on the other hand, who thinks of humans as evolved organisms first and foremost, might say that autocracies are best explained as one expected product of dominance relations, which are commonly found in other primate societies and hypothesized to have existed in ancestral species. Here considerations of individual rationality don't apply.

The biologist's explanation may appear to be killjoy compared to the social scientist's romantic description of the beliefs – and the intrigue – that play a role in their explanation. But both can be apt for the context in which they are created. And both can be true.

Daniel Dennett introduced the terms "killjoy" and "romantic" into the lexicon of comparative psychologists. He describes the "killjoy bottom of the barrel [as] an account that attributes no mentality, no intelligence, no communication, no intentionality" and "the most killjoy [as the] least romantic hypothesis" (Dennett 1983, 346).

These terms "killjoy" and "romantic" haunt discussion and debates about animal mental processes, but, like worries about anthropomorphism and Morgan's Canon, they do so without promoting the fecundity and maturity of comparative psychology. While the debate is unhelpful, diagnosing its cause is informative – it is grounded in a false shared assumption that there is a single science of animal minds or that there is some objective position from which we will be able to adjudicate between disciplines to find the one true explanation of animal behavior. As participants in a multidisciplinary endeavor, scientists use various tools and concepts in order to investigate what appears to be the same question. Because the question is framed by the discipline that surrounds it, the kinds of answers the scientists are looking for will differ, and they need not be in conflict any more than are the social scientist's and the biologist's explanations of the same human behavior.

Work in the philosophy of science on objectivity and bias makes it clear that there is no objective "view from nowhere" from which we can answer a question like whether the biologist's or the social scientist's explanation of the autocrat's rise to power is better or more accurate (e.g., Longino 1990). And there is no view from nowhere from which we can answer a question such as whether an associative or cognitive explanation of some animal behavior is better. Rather, all the answers should be interpreted from the context in which they are asked and given the goals of the scientists who are asking them.

3.2 A Middle between Romantic and Killjoy?

In 2012 the Royal Society held a meeting, "Animal Minds: From Computation to Evolution," dedicated to exploring how animals engage in behaviors that look like behaviors once thought to be uniquely human – behaviors such as hierarchical tool use, mourning the dead, empathy in rats, self-recognition, and so on. The fifteen listed speakers were top scholars from eight different disciplinary backgrounds: artificial intelligence, biology, computer science, ecology, neuroscience, psychology, philosophy, and zoology. In his report on the conference for *Science* titled "'Killjoys' challenge claims of clever animals," journalist Michael Balter describes a tension in the field between those who see continuity between animals and those who see discontinuity.

Balter talked to Daniel Dennett, who told *Science*, "People in the field often gravitate into two camps. There are the romantics," those who are quick to see humanlike traits in animals, "and the killjoys," who prefer more behaviorist explanations. "I think the truth is almost always in the middle" (Balter 2012, 1036).

Is it helpful to take Dennett's claim about the truth being somewhere in the middle literally? I don't think so. To consider the possibility, we can return to Dennett's introduction of the distinction in the context of vervet monkey alarm calls. With the publication of Dorothy Cheney and Robert Seyfarth's book *How Monkeys See the World* (1990), there was much interest in the discovery that vervet monkeys give different alarm calls for different predators (Seyfarth et al. 1980). (We now know that many species of animals, including domestic chickens, give different alarm calls for different predators.) Dennett introduced a distinction between higher levels of intentionality (romantic) and lower levels (or zero level) of intentionality:

> We can now compose a set of competing intentional interpretations of this behavior, ordered from high to low, from romantic to killjoy. Here is a (relatively) romantic hypothesis . . .
>
> *4th-order:* Tom *wants* Sam to *recognize* that Tom *wants* Sam to *believe* that there is a leopard . . .
>
> *3rd-order:* Tom *wants* Sam to *believe* that Tom *wants* Sam to run into the trees . . .
>
> *2nd-order:* Tom *wants* Sam to *believe* that there is a leopard.
>
> *1st-order:* Tom *wants* to cause Sam to run into the trees (and he has this noise-making trick that produces that effect; he uses the trick to induce a certain response in Sam) . . .
>
> *0-order:* Tom (like other vervet monkeys) is prone to three flavors of anxiety or arousal: leopard anxiety, eagle anxiety, and snake anxiety. Each

has its characteristic symptomatic vocalization. The effects on others of these vocalizations have a happy trend, but it is all just tropism, in both utterer and audience.

We have reached the killjoy bottom of the barrel: an account that attributes no mentality, no intelligence, no communication, no intentionality at all to the vervet. (Dennett 1983, 346–47)

Unfortunately, going back to the original text doesn't help us understand the location of the truth "in the middle," for Dennett surely doesn't mean that the truth would be at the 2nd-order (between 4th-order and 0-order). Rather, I think it suggests that we need to look at the issue in a different way.

Dennett's "the truth is in the middle" suggests that there is a single field with two camps – an anthropomorphic and a behaviorist camp – and some people see animals anthropomorphically and others see them behavioristically. But comparative psychology is not a single field of investigation. Compared to the study of human minds, there are many fewer scholars teaching and researching in the area and many more species to talk about. In comparative psychology, interdisciplinary conversations are almost inevitable, and disagreements will arise due to different disciplinary foci and methods.

Dennett's levels of intentionality might look like explanatory competitors, but they are not. Vervets, like humans, could have automatic responses to vocalize when they see danger approaching, implementing something like a 0-order scheme. But they might *also* want to scare away the intruder or protect their companions (1st-order), and they might also want to change their companion's epistemic states (2nd- to 4th-orders). Humans are complex, and vervets might be just as complex. We have mixed motives. We have automatic responses. These orders of intentionality do not present an exclusive disjunction of possibilities.

There is no continuum between behavioristic and anthropomorphic explanations; there is no middle between them where some truth resides. Rather, explanations in terms of associative mechanisms and explanations in terms of concepts, culture, or mental states can be consistent. This is as true of humans as it is of other animals.

3.3 Avoiding Killjoy and Romantic with a View from Nowhere?

We might find the answer of what lies between romantics and killjoys if we turn to another description of the terms. Elliott Sober offers a kind of sociological speculation as to why some think romantic mistakes are worse than killjoy mistakes when he writes, "mistaken anthropomorphism is often taken to reflect a kind of tenderheartedness, whereas the . . . error of mistaken anthropodenial is supposed to reveal a kind of tough-mindedness" (Sober 2005, 86). Tenderheartedness and tough-mindedness are two kinds of perspectives, and so to find the middle we want to find another perspective. What would be the middle between these two perspectives? Objectivity, or the "view from nowhere" – an unmediated conception of the objective features of the world – reality that isn't described from a particular contingent perspective and that isn't warped by human perceptual processes.

Of course, we can't do science without being engaged in perception, so the view from nowhere at best can serve as an ideal to aim for. But at worst it is an aim that eliminates the varieties of knowledge available from various perspectives and thus results in a loss of information about the world.

Objectivity is a virtue of science, but philosophers and scientists disagree about what objectivity is. In a paper in *Nature*, Mary Midgley criticizes those scientists who attempt to be objective by prohibiting what Evelyn Fox Keller calls "a feeling for the organism" – leading to relationships between scientists and those beings they study (Midgley 2001). Looking at the development of behaviorism in the early 1900s, Midgley directs us to consider John Watson's insistence that scientists be objective in the sense of being free from emotion. Not having emotional attachment is particularly important in the case of children, Watson thought, and not just for scientists. In his best-selling guide on raising children, Watson advised parents:

> There is a sensible way of treating children. Treat them as though they were young adults … Let your behaviour always be objective and kindly firm. Never hug and kiss them, never let them sit on your lap. If you must, kiss them once on the forehead when they say good night. Shake hands with them in the morning. (*Psychological Care of Infant and Child* 9–10; W. W. Norton, New York, 1928)

This prohibition against becoming attached, Midgley thinks, explains Jane Goodall's problems getting her early work published without editorial interference. Goodall's editors sent back manuscripts, correcting her prose so that chimpanzees became "its" rather than "he" or "she."[2] This early debate about pronouns was one that Goodall ended up winning. She named the chimpanzees she observed and worked with, and she considered them to be beings worthy of agential pronouns, because she formed relationships with the chimpanzees as individuals. She got to know them.

It was for just these reasons that other scientists dismissed her work as unscientific. As the Anti-anthropomorphic principle directs, scientists should not develop a relationship with their subjects. The extent of that relationship,

---

[2] Goodall's problems with editors harkens back to our discussion in Section 1 of Monica Gagliano's problem getting her plant studies published, not because editors had a problem with her scientific methods or analysis, but because they rejected her use of words like "learning" and "intelligence."

which included babysitting infant chimpanzees and playing with grown individuals, is apparent in the recent National Geographic documentary *Jane*, which uses newly discovered archival film taken by Goodall's ex-husband Hugo van Lawick. What is particularly shocking about this film is how familiar Goodall is with her subjects compared to "best practices" in current field research. Goodall is shown playing and wrestling with adult chimpanzees and cuddling infants. Current guidelines forbid physical contact with wild apes because they are susceptible to human diseases (something we were not as aware of in the 1960s). While cuddling baby apes should generally be avoided, forming relationships that are safe for scientist and ape alike has turned out to be key to producing good research.

The entreaty to respond to children and animals without emotion, Midgley argues, is not the same as calling for objectivity in science. Rather, reacting to others without emotion introduces a different kind of bias, as it treats others "as a lifeless object, not as a subject" – an "it," not a "he" or "she" or "they." This treatment limits the kinds of questions we can ask and the kinds of things we can see. Furthermore, as we saw, the relationship between scientist and subject is a relevant variable (e.g., friendly vs. frightened handling of a rat subject) and may be a scaffold rather than a cue (e.g., imitating a caregiver but not a stranger).

Indeed, Midgley argues that the goal of objectivity should follow our ordinary sense of the word, "which is simply *fair, unbiased, impartial*" (Midgley 2001, 753). While we may be able to engage in *fair* science to a certain extent, applying the same kinds of tests and requirements appropriately to different subjects, I'm pessimistic about our ability to engage in *unbiased* or *impartial* science. Bias is inherent in how we see the world, and, I'll argue, the best way to deal with bias is to identify it where it exists.

Numerous worries have been raised about getting to the view from nowhere, and we don't need to review them here. There will be bias; there is always partiality. The goal might be to minimize bias, but the best way forward might be to use multiple approaches. At any rate, the middle cannot be a no-bias standpoint, a no-partiality view, since science is a human endeavor and humans have perspectives.

Midgley's suggestion that objective means fair, unbiased, and impartial undermines the very possibility that we can reach objectivity. As there will always be uncertainty in our scientific judgments, we can leave aside that sort of objectivity as an achievable goal and instead acknowledge that bias exists. In that way, romantics and killjoys can be seen as two perspectives, each with its own biases. This view is helpful, because it allows us to examine the kinds of biases that come with each of these perspectives.

3.4 Sidestepping Romantic and Killjoy?

In a *Trends in Cognitive Science* opinion article, Sara Shettleworth argues that scientists should retire the killjoy and romantic concepts. She writes, "in the contemporary study of animal cognition, demonstrations that complex human-like behavior arises from simple mechanisms rather than from 'higher' processes, such as insight or theory of mind, are often seen as uninteresting and 'killjoy', almost a denial of mental continuity between other species and humans" (Shettleworth 2010a, 477). The field and the media's tendency to get more excited about romantic mechanisms than simple mechanisms is, Shettleworth thinks, a problem that gets in the way of a true comparative approach to cognition. High-level explanations are exciting (and will get published in *Science*). Low-level explanations are boring and will have a more difficult time getting published. This leads to a bias in the science that is detrimental.

I think Shettleworth is largely correct on this point. The animal cognition papers that appear in *Science* and *Nature* tend toward reporting findings that animals can do something interesting, or something that was thought to be unique to humans. Take the ape theory of mind research program, for example. The first paper on this topic appeared in *Behavioural and Brain Sciences* in 1978 (Premack and Woodruff 1978). Forty years later, *Science* published the first positive report of apes passing a false belief task (Krupenye et al. 2016). Between the publications of those two papers, null results were all published in more specialized journals. It took passing the false belief task to get published in *Science*. Shettleworth suggests that this trend is due to an unexamined idea that higher-level explanations support continuity claims, and continuity claims are exciting. Lower-level claims, on the other hand, are felt to be evidence of discontinuity. Shettleworth thinks this viewpoint gets things just wrong.

When we look at human psychology, we see a trend toward looking at the automatic, associative, and unconscious processes that drive much of human behavior. Shettleworth writes, "The tendency in comparative cognition to emphasize the human-like in animals is curiously out of step with an important trend in cognitive and social psychology toward uncovering what is essentially the animal-like in humans" (Shettleworth 2010, 479). Continuity is, well, a continuum. Continuity doesn't have a directionality. If A – B – C is a continuous series, then all things being equal, it is as likely we'll see A-properties in B as we'll see B-properties in A.

The recent rise of interest in seeing what is animal-like in humans is apparent across psychology, from the confabulation literature in social psychology to the dual-systems literature in cognitive psychology. But there has been little interest in describing these explanations as killjoy. For example, the confabulation research suggests that humans sometimes adopt false reasons for their own actions. In a landmark study, shoppers were asked to select the "best" pair of pantyhose from an array (Nisbett and Wilson 1977). The majority of subjects strongly preferred the rightmost pantyhose and, when asked to explain their preference, they offered reasons – this pair is softest or has the best color. But since the items were identical, the shoppers were wrong about the cause of their action. The hypothesis for why subjects chose the rightmost item is that because the subjects read from left to right, they have a positive association with items at the end of an array that mimics their writing practices. If this explanation is accurate, we have an associative explanation for the subjects' actions.

Morgan himself recognized that much human cognition is animal-like. In his biography, he wrote, "To interpret animal behavior one must learn also to see one's own mentality at levels of development much lower than one's top-level of reflective self-consciousness. It is not easy, and savors somewhat of paradox" (Morgan 1930, 250). This methodological principle, which I've been calling Morgan's Challenge (Andrews 2014), tells us to be as careful thinking about the causes of human behavior as we are in thinking about the causes of animal behavior. Here Morgan is anticipating not behaviorism but much of the recent work in cognition on the role of heuristics and biases, dual-process models of cognition, and the embodied mind.

Morgan's Challenge – the idea that there are good "killjoy" explanations of human behavior too – is gaining more acceptance by philosophers and animal cognition researchers. Cameron Buckner (2013) and Louise Barrett (2011) both point to a problem in animal mind research that arises from a false view about human minds, which is what Morgan's Challenge warns us against. Buckner identifies a methodological error in animal mind sciences that he dubs *anthropofabulation* – attributing superhuman cognitive capacities to humans and using this exaggerated description of human competence to test for the same competence in other animals (Buckner 2013).

Of course, it would be an error in reasoning to identify some property as a property of humans *when humans don't actually have that property*, look for that property in nonhuman animals, and when that property isn't found conclude that mental continuity does not exist across species. But is noting this error, and noting that one is just as likely to find chimpanzee properties in humans as one is to find human properties in chimpanzees, enough to sidestep the debate between romantics and killjoys?

It should be, but there is another connotation of "romantic." A romantic takes a special human property, one that has been deemed unique to humans – such as mourning the dead, having friends, or following moral norms – and finds some aspect of that property in other animals. That is, an anthropectic background assumption regarding human uniqueness is threatened by the romantic. There is more than a whiff of human superiority in the grouping of behaviors and capacities as "high" rather than "low" or "romantic" rather than "killjoy."

When Daniel Povinelli worries that scientists are biased when they look for similarities across species, he appears to be worried that some supposedly special human property will be inappropriately ascribed to humans. If we are expecting to find similarities, he thinks, then we will find them, and we potentially will miss some possible "qualitatively new cognitive systems" that emerged in the *sapiens* lineage during the last 4 million years (Povinelli and Bering 2002). Looking for similarities, Povinelli warns, means we won't be able to uncover what is unique about humans – or what is unique about chimpanzees either.

In response to this kind of worry, we can note that if Povinelli is right and looking for similarities is biased, then so is looking for differences; if we are expecting to find differences, then we will find them, and we will miss the possible continuities that connect the *Homo* lineage with the *Pan* lineage. The disciplines of comparative psychology, which all seek to find similarities *and* differences between species, would be biased as well. But remember, the mere identification of bias isn't a criticism of the science. Rather, identifying bias can be a helpful exercise in the development of improved scientific metamethodologies for interdisciplinary research.

3.5 Locations of Bias

Dennett, Povinelli, Sober, and Shettleworth are all concerned about the role of bias in animal mind sciences. What I hope will be helpful is to identify where we might find biases in the sciences of animal minds and what those biases are. Noticing the biases in a scientific practice is particularly important when engaged in multidisciplinary research, given that the biases will likely differ given different methods. With an articulation of what those biases are, we can form better overhypotheses that explain and unify a body of findings about a species, especially when those findings come from different disciplines. We can identify three kinds of bias in the sciences of animal minds: in choice of measurement systems, in choice of theoretical terms, and in what counts as the proper topics of investigation.

The measurement systems we use in science, whether physics, chemistry, or psychology, tell us about our subject of study as well as about the instruments being used and the goals of the people who designed the instruments (Giere 2006). The telescopes of astronomy and the personality assessments of psychology each have assumptions built into them, as does the comparative psychologist's ethogram. The outputs of these instruments are themselves subject to further interpretation, and these interpretations are going to be given in light of other aspects of the relevant science, including the accepted theoretical terms and views about what counts as proper areas of investigation. Measurements are theory laden and take as given particular metaphysical and theoretical perspectives.

The theoretical terms we use in science can be standardized and operationalized, but our choices in how to do so shape the scope of investigation. For example, we might define "perception" as the representation of sensory stimuli and operationalize perception in terms of demonstrating perceptual constancies. Such a term is both theory laden and involves hypotheses about how to test for the postulated representational capacity. When we turn to terms such as empathy, we have less evidence to support one account over another. For example, if "empathy" is operationalized as the ability to accurately report what another person is thinking as in the "empathic accuracy" research, then empathy in nonverbal individuals will be ruled out by definition.

The topics of investigation comparative psychologists choose to pursue, such as the study of personality, unconscious processing, stereotype threat, conformity bias, active perception, or priming, reflect human interests. Science is not value free and only directed at truth for truth's sake, since there are infinite truths that we don't seek at all (Longino 1990). Science seeks out truths we care about, truths that are significant to us. For example, not investigating social norms in animals may reflect human disinterest in the topic, or perhaps the fear we'll discover humans are not the only species who have rules to live by.

Measurement systems, theoretical terms, and topics of investigation come together in many cases to shape how scientists choose to do science. One case in point is the work on animal culture, which today is a hot topic. Animals including great apes, cetaceans, canids, rodents, birds, fish, and insects have all been reported as having culture (for a review, see Allen 2019). However, describing animal traditions as "culture" was nearly taboo in North America until 1999. That is the year that a group of scientists, representing seven different chimpanzee communities across Africa, found differences in behaviors that were not attributable to genetic or ecological factors (Whiten et al. 1999). Recognizing animal culture required an expansion of the definition so that it could include animals. Whiten and colleagues point out that cultural anthropologists often define culture as requiring linguistic transmission, and that this anthropocentric definition is unhelpful for a comparative investigation.

They adopt a more inclusive definition of culture that comes from prior etho-logical work on animal traditions: "a cultural behaviour is one that is transmitted repeatedly through social or observational learning to become a population-level characteristic" (Whiten et al. 1999, 682). There are ongoing discussions about how best to define culture in a way that captures what is key to human culture without capturing what is contingent to any particular taxon or species (Avital and Jablonka 2000; Laland and Janik 2006; Ramsey 2017; Rendell and Whitehead 2001).

While the animal culture conversation is relatively new in the West, more than fifty years ago in Japan, the primatologist Kinji Imanishi described the culture of a community of Japanese macaques on Koshima Island. These macaques wash sweet potatoes in water before eating them (Imanishi 1957). Imanishi and his colleagues' observation of the unusual washing behavior led them to look for additional differences between various macaque communities in Japan. They discovered that there are not just differences in food processing behaviors across macaque communities, such as the food-washing, but also differences in social behaviors. When Imanishi traveled to the United States in 1958 to report on their discoveries of culture in other animals, he was widely derided (de Waal 2003). The Western scientists weren't criticizing Imanishi's anti-Darwinian evolutionary theory but rather his rejection of Anti-anthropomorphism and his descriptions of methodology. For example, scientists openly expressed disbelief that Japanese scientists were able to recognize individual monkeys. When the first Japanese chimpanzee field site was established in 1965 by Toshisada Nishida in the Mahle mountains of Tanzania, researchers adopted an anthropological approach and examined the "species society" – the relationships between all members of the group (Asquith 1996). Nishida, the academic grandson of Imanishi, did not receive the same kind of criticism back in Japan that Jane Goodall had to confront in her early years due to her "unscientific" methods of describing chimpanzees as agents. Differences in how primatology findings were reported in Japan and the West continued through at least the 1960s (Asquith 1996).

The disagreement between Japanese and Western scientists about animal culture reflects a difference in how they identified culture and a difference in the willingness to ask the question; measurement systems, theoretical terms, and topics were all in dispute. De Waal describes how Imanishi wondered about animal culture years before the sweet potato washing behavior was observed:

> As far back as 1952, when European ethologists were working on instinct theories and American behaviorists were rewarding rats for pressing levers,

Imanishi wrote a paper that criticized established views of animals (Imanishi 1941, 1952). He inserted a debate between a wasp, a monkey, an evolutionist, and a layman, in which the possibility was raised that animals other than ourselves might have culture. Hirata et al. (2001) provide a translation of a portion of this imaginary debate. The proposed definition was simple: if individuals learn from one another, their behavior may, over time, become different from that in other groups, thus creating a characteristic culture (Itani and Nishimura 1973; Nishida 1987). (De Waal 2003, 296)

With a proposal for how to define and investigate animal culture, and the acceptance of it as a proper question for the science, Imanishi and his colleagues were able to find it in the monkeys and apes they studied.

Today's definitions of culture vary and do so in accordance with the disciplines, goals, and theoretical backgrounds of researchers. In Kevin Laland and William Hoppitt's (2003) review of many diverse definitions of culture, two points of consensus emerge. First, culture consists of information that is socially learned and transmitted, not transmitted genetically or learned on one's own. Second, culture is specific to a population or group, and underpins group-typical behavioral patterns which can help to explain conformity within groups and diversity between groups.

Imanishi's definition of culture is not that different from the definition of culture that emerged from Laland and Hoppitt's review: "Cultures are those group-typical behavior patterns shared by members of a community that rely on socially learned and transmitted information" (Laland and Hoppitt 2003). It took only sixty years for Western science to endorse an Imanishi-style definition of culture.

3.6 Sources of Bias in Romantic and Killjoy

In this section I aimed to show that different kinds of bias will result from different scientific approaches, be the differences disciplinary or cultural. A romantic bias isn't worse than a killjoy bias, if the goal is to get at the truth. What helps promote that goal is a recognition of sources of bias. Once we identify that bias can arise in measuring systems, theoretical terms, and topics of investigation, we can put those biases on the table when evaluating our current state of knowledge on a topic and when investigating what sorts of questions remain to be investigated. The biases will always be with us, but acknowledging them can help to increase our understanding by seeking evidence from different disciplines and by including scientists from different cultural backgrounds. In the next section I will illustrate the point by examining how the romantics and killjoys debate plays out in ape studies, and suggest a path forward.

4 Biases in Ape Cognition Studies

4.1 All about Apes

The great apes – chimpanzees, bonobos, gorillas, orangutans, and humans – are the focus of much comparative research in animal cognition, and the methods and results from the science are perhaps among the most disputed. There is an ongoing tension between captive research on great apes, where experiments can be designed using standard controls, and field research on great apes, where observations can identify social structures and cultural practices. In this section I aim to offer a reconciliation between the two approaches – field and lab – by identifying the biases apparent in each approach.

I will not defend one approach over the other; rather, my goal is to show how fieldwork and lab work are complementary. By identifying the biases in both approaches, we can reach better answers to our scientific questions. To cultivate the kind of epistemic humility that allows scientists to see the limits of their approach is to set the foundation for productive interdisciplinary research that can provide us with a much richer understanding of other animals.

To begin, I will note that even as comparative psychology is multidisciplinary, ape cognition research is more so, including researchers working in psychology, philosophy, biology, anthropology, and primatology departments. Field researchers often require expertise in issues related to climate, botany, endocrinology, and geology. The different disciplines use different instruments, different observational terms, and work on different general topics. Field researchers and experimentalists tend to go to different conferences, meeting only at larger primate meetings where it is easy to observe groups of field researchers still wearing their Columbia Sportswear gear in one session and lab researchers garbed in standard academic wear in another. They also conduct their research in different places and with different kinds of subjects. Anthropologists tend to work with wild animals in the field, while psychologists tend to work with captive animals in labs or zoos. This is by no means a universal claim – especially given the recent trend of captive researchers travelling to sanctuaries in Africa and Indonesia to conduct (mostly experimental) research – but it is an accurate generalization. I will also be drawing a rough distinction between "captive research programs" and "field research programs." These also oversimplify, as standard field methods can be used in some captive settings, such as ethograms used in zoos, and captive methods can be used in some field settings, such as playback experiments in the field. With these caveats we can turn to examine the differences in measurement systems, choice of theoretical terms, and topics of investigation. Then we can turn to examine the debates in the literature that have arisen about the relative merits of research methods used with wild and captive apes. I will present a list of the sources of bias in both fields with the goal of showing how the two kinds of research can each offer a rich source of knowledge about ape minds.

4.2 Field and Lab, Romantic and Killjoy

The categories of field and lab research correspond roughly with romantic and killjoy perspectives – or at least they correspond with being a likely target of such an accusation. Field researchers, who tend to have more of a biological background, as well as captive researchers who use observational methods, are more often taken to be romantics – take Christophe Boesch, Jane Goodall, and Frans de Waal. Captive researchers, who tend to have more of a psychology background, are more often seen as killjoys – for example, Cecilia Heyes, Michael Tomasello, Daniel Povinelli, and Sara Shettleworth.

In order to illustrate how this conflict can play out in ape research, we can look at the responses to a field study I coauthored (Andrews and Russon 2010). While working with ex-captive juvenile orangutans under rehabilitation to enable their return to free forest life in Indonesian Borneo, psychologist Anne Russon and I noticed that the orangutans would sometimes act out what they wanted their human caregivers to do for them. Before we could even recognize the behaviors as requests, we had to learn about this orangutan–human community. We spent time with these orangutans and their human caregivers and learned what their typical practices and behaviors were. It was easy to see that the orangutans regularly gathered together in dusty areas, wrestling in the dirt, collecting handfuls of dust like children in a sandbox, and dumping it on their own heads. We also soon came to expect that their human caregivers would clean the little orangutans after their play bouts, brushing the dirt from their heads with leaves. Given our observation of normal behavior in this community, we were able to recognize a behavior as a request from an orangutan, Cecep, for Russon to clean him. Cecep approached and sat in front of Russon, picked up a leaf, and handed it to her. Russon used it to briefly clean Cecep's head, then dropped it on the ground. Cecep picked up and handed Russon another leaf, but this time she played dumb and just examined the leaf. After a few seconds Cecep took the leaf back from Russon, rubbed it on his own head while looking her in the eye, and then placed it on her notebook. Then Russon picked up the leaf and actively cleaned Cecep's head. We interpreted this event as Cecep asking Russon to clean his head by handing her the leaf, and when she didn't respond as he expected, Cecep elaborated on his message by pantomiming – acting out – what he wanted Russon to do.

This incident led Russon to recall that she had witnessed and recorded similar behaviors during her previous twenty years studying rehabilitant orangutans. Like other field researchers interested in studying unusual behaviors in animals, we used a data mining technique, identifying eighteen reports of orangutan communicative behaviors that qualified as pantomime. (Subsequently, Russon (2018) updated the dataset with 62 pantomimes.) With this dataset, we were able to analyze the contexts in which the pantomimes were exhibited in order to determine the functions of these gestures. We found that in all but one case the orangutans used the gesture imperatively; they tended to use it to elaborate a prior failed message. In seven cases they used pantomime in a deceptive context, and in one case an orangutan pantomimed in a declarative context.

When this study was published, it received a bit of media attention, with articles in news magazines and reports on the radio. This study was quickly placed into the romantic category, and journalists sought out kill-joys to create some conflict, as journalists like to do. *Science News* contacted Michael Tomasello for his opinion on the study, and they quote him as saying, "Without some kind of control observations we cannot be sure what [the orangutans] are doing ... How often do the orangutans make those hand movements in other, irrelevant contexts?" (Milius 2010).

In this quotation, the methods of the lab, namely controlled repeatable behaviors elicited by the same kind of stimuli, are inappropriately imported into field research. It isn't clear how a field researcher could provide a relevant kind of control for pantomime communication. We could observe Cecep for some number of hours and report all the situations in which he rubs his head with a leaf and then hands it to someone when his head isn't dirty – but even if he did behave that way, he may be asking for a head massage rather than a cleaning. We could report all the times Cecep handed a leaf to empty air rather than a communicative partner, but we don't need to do that, because Cecep wasn't observed to treat empty space communicatively. Rather, we saw the gesture as communicative because we spent time enough with Cecep to know how to engage with him; we had a relationship with Cecep. (It wasn't many days after the pantomime that Cecep offered me both a flower and his erection.) Russon's twenty years of experience gave her the skill of knowing how to coordinate with orangutans, to communicate with them in such a way so as to preserve the consistency in interaction, as one would expect from a successful application of the principle of charity. Russon's methods in this case resemble anthropology much more so than psychology. Linguistic anthropologists begin interpreting communicative signals made by humans in other groups, and as their interpretations permit greater predictive power and coordination with others, anthropologists gain greater confidence in their interpretation.

Of course, worries of radical interpretation emerge when it comes to communicating with other species as well as members of other cultures. But we cannot forget that it also exists when it comes to communicating within one's own linguistic community. The best we can do with other humans and other species is to adopt a principle of charity so that others' actions make rational sense. The principle of charity as applied to Quinian radical translation has always been a methodological principle – in humans, we presume that our interlocutors are rational and that they have mostly true beliefs in order to figure out what they mean by their utterances. Using the principle of charity to understand other animals is also a gambit – we presume that the animal is rational and has mostly true beliefs, and in light of the gambit form expectations about what the animal should do in particular circumstances. The gambit is warranted as a strategy given an approach that takes animals as conscious agents. Consciousness isn't sufficient for being a rational agent, but judging someone as conscious certainly does raise the probability that we'll also take them to be rational. Since the principle of charity is just a methodological principle in this context, if the predictions robustly fail to bear out, we reject the gambit.

While one may object that humans share an evolutionary history and a linguistic practice that we don't share with apes, making interpretation more difficult in the case of apes, we shouldn't forget that human linguistic practice evolved from simpler communication systems that also were subject to interpretation. Groups create communicative systems by treating one another as agents with content to communicate, and will stop doing so only if that treatment doesn't work – that is, if the interpretation doesn't promote prediction or coordination. Even if a shared language reduces the degree of uncertainty in interpretation to some extent, there still remains an interpretive task, and robust patterns of prediction provide the best evidence. As we will see, Russon has been quite successful in interpreting the orangutan, across contexts, in ways that work.

When Kuhn wrote, "Practicing in different worlds, the two groups of scientists see different things when they look from the same point in the same direction" (Kuhn 1962) he could have been talking about field and lab ape researchers. In the orangutan pantomime case, the different worlds are the perspectives from which the scientists look at the human–orangutan society. The caregivers and field researchers have a perspective from within, based on extensive shared experiences while spending their days with their subjects and cocreating community with them. The lab researchers who criticized the study have a perspective from outside the community and missed some of the information due to their more distanced, but less informed, epistemic stance.

Our pantomime study took what anthropologists call an emic approach, and the criticism came from an etic approach. These concepts, coined by anthropologist Kenneth Pike in 1954, provide a helpful strategy for animal mind scientists. Emic approaches to studying culture approach the culture from the inside, such that the scientist enters the group and becomes part of the community under investigation. The term is also used to refer to a more other-regarding interpretation of actions, events, and objects in a community. Etic approaches to studying culture approach the community from the outside and attempt to use the observer's own concepts and categories when describing the actions, events, and objects that are observed. Anthropologists have spilled much ink on how to walk the tightrope between etic and emic approaches, and syntheses of emic/etic approaches also have been adopted. One weakness of the etic approach is that it may be unsuitable for describing the phenomenon of interest if the external perspective includes irrelevant concepts, such as personality measures that do not share the same factors across cultures (Coulacoglou and Saklofske 2017). This weakness is a strength of the emic approach, which seeks to discover the concepts that are relevant to the community at hand. When we use a descriptive system that is validated for all cultures, we may miss the differences and only categorize the similarities. A strength of the etic approach is the ability to build large databases of performance across cultures and to make comparisons using the same construct. This strength corresponds to a weakness of the emic approach, which does not permit such comparisons. Researchers who want to understand both the local and the universal features synthesize the approaches, as demonstrated in the cross-cultural personality research referenced above. Primatology, as the study of primate cognition, behavior, and culture, could likewise benefit from a synthesis approach.

4.3 Debates about Bias in Ape Cognition

The biases inherent in field research have been identified by lab researchers, and the biases in lab research have been identified by field researchers. Like recognizing the norms of one's own culture, it can be easier to see biases from an outsider's perspective. However, this doesn't make it any easier to hear the critiques. I hope to offer an analysis of these biases from an informed outsider's perspective. My own experiences working on artificial language and memory studies with captive dolphins at Kewalo Basin Marine Mammal Laboratory in the 1990s and with rehabilitant and wild orangutans in Borneo in the 2000s–2010s has given me some firsthand experience with both kinds of methods without leading me to be entrenched in either approach.

Table 4.1 lists the potential sources of bias in field and captive research, which will be discussed in some detail. To begin, however, I will compare the biases that can arise in measuring systems, theoretical terms, and topics of investigation in field and captive studies. With those initial sources of bias on the table, we will then dig a bit deeper to see other places in which bias may emerge in both methodologies.

Table 4.1 Sources of Bias in Field and Captive Ape Studies

Potential sources of bias in field research	Potential sources of bias in captive research
Measurement systems	Measurement systems
• Ethograms limit what is seen	• Data sheets limit what is seen
• Ethograms require interpretation	• Little opportunity to record qualita-
• Ethograms might encourage confir-	tive description (e.g., on motivation
mation bias	or interest)
• Video does not capture context of	• Relationships between individuals
actions off screen	not standardly recorded
• Camera traps only record what hap-	
pens in certain locations	
• Narrative descriptions are qualitative	
Observational terms	Observational terms
• Operationalized (e.g., peering)	• Operationalized
• Ill-defined (e.g., grieving)	• Theoretically defined
Topics	Topics
• Anthropological, social, and cul-	• Community composition can limit
tural intergenerational investiga-	anthropological, social, or cultural
tions possible	investigation
• Cognition may be difficult to study	• Suitable for examining possible
• Data are largely limited to activity	cognitive mechanisms supporting
perceptible by humans (e.g., day-	behavior
time, not up too high, not in deep	
brush, not behind branches, not	
vocalizations outside the normal	
human range)	
Context	Context
• Observation effects	• Controls are available
• Incomplete habituation of subjects	• Physical barriers between experi-
• Lack of controlled environment	menter and participant
• Limited repeatable events	• Tested by contraspecifics
• Research area and camp location	• Use of species-relevant materials
choices	• Use of species-relevant social sys-
	tems, communication systems, and
	norms
	• Proximity to social support during
	testing

Table 4.1 (cont.)

Potential sources of bias in field research	Potential sources of bias in captive research
	• Relationships (between experimenter and subject; between conspecifics)
	• Ecological validity
	• Quality and nature of enrichment
	• Quality and nature of social groups
	• Quality of experimenter's knowledge about the species

4.3.1 Measurement Systems, Observational Terms, and Topics of Investigation in Field and Lab

The typical measurement system for field researchers is the ethogram. An ethogram is a catalogue of behaviors used to report how frequently and in which contexts some set of behaviors are observed, and it is usually carried into the field as a datasheet on a clipboard or a handheld computer. For example, in a field study of orangutans that uses a focal sampling method, a researcher may observe an individual from "nest to nest" (that is, morning to night) and record on the datasheet each instance of a behavior listed on the ethogram, or they may record behaviors at regular intervals – say every five minutes. These behavior types might include travel, feeding, nesting, peering (i.e., observation of another's actions from less than one meter away), and other social behaviors. Ethograms usually also include space to record information about the target of the behavior (e.g., how one is traveling or what one is eating), and other proximate individuals. The behaviors listed on the ethogram are carefully defined or operationalized.

Colin Allen and Marc Bekoff describe a decision that has to be made in constructing an ethogram, which presents another source for bias. Researchers can choose to describe behavior *functionally*, in terms of its purpose, or *formally*, in terms of the actual movements of the body (Allen and Bekoff 1997). A functional item on an ethogram would be "social play," whereas a formal item on an ethogram that hopes to capture the same behavior could be put in terms of physical proximity and contact with another animal. Allen and Bekoff note that formal descriptions can miss important aspects of an animal's behavior; play and fight will look the same on an ethogram that measures only physical proximity and contact. Likewise, functional descriptions can misinterpret animal behavior; play and consort may look the same to an observer who has not yet observed the species' sexual behavior. Furthermore, if the term "play" has a rich connotation to the researcher, they may draw inferences that are unwarranted, including inferences about the relationship between the play partners or their affective states.

An ethogram is limited because it restricts the researcher to record the sorts of behaviors and variables that appear on the ethogram – those that were deemed, prior to data collection, to be relevant to the study at hand. New insights that arise in the course of data collection cannot be integrated into the current study using standard methods. In addition, items may be missed if terms are defined too precisely. For example, if "peering" is defined as close observation of another's action from less than one meter away, then an individual won't be recorded as peering if they observe from 1.25 meters. Instead, the behavior might be recorded as another behavior, such as a "look." A "peer" data point might be lost, along with the opportunity to gain insight into whether social learning can occur between individuals who do not tolerate one another well. These limitations have been minimized by introducing new methodologies that include rules of thumb such as "also record any interesting or unusual incident" using a "scribble method" in which narrative descriptions of ape behavior are also continuously recorded, or via video recordings that can be analyzed later. Such descriptions may be hard to code, and require interpretation at a later stage of the investigation when doing analysis, since the terms used may be ill-defined.

Ethograms may encourage confirmation bias by priming field assistants to see just those behaviors that are on the ethogram. Unlike captive tests, in which the experimenter can at least sometimes be ignorant of the correct answer, with ethograms field assistants have to know what they are looking for. If a field assistant knows that the primary investigator is studying deception, then it may prime them to see the "deception" pattern of behavior where they wouldn't have otherwise. It might also keep them from seeing other behaviors that are not in the ethogram. In addition, researchers might throw out data that were collected when a focal subject appears to be ill, lazy, or otherwise acts atypically, introducing another place for a judgment call.

Field researchers might attempt to overcome the limitations of ethograms by video recording the behaviors that are being coded on the ethogram. The video evidence can later be used by independent observers, but this requires having good videographers and cameras, which creates new challenges. Videos also fail to indicate to their audience what was happening in the larger context, off scene.

The typical measurement system for captive studies is a data sheet that is used to indicate how the animal performs in repeated trials of an experiment. Pass and fail criteria are clearly defined prior to the start of the study, and if an animal makes a "rational" mistake, it may still count as a fail. Other information, such as judgments about the animal's motivational state or mood, can be indicated on the data sheet, but such information does not usually make it into analysis of the data, which is limited to the coded performance on the trials. Information about motivation can impact the final results if a research team decides to drop a subject for some reason, such as not being motivated. Other variables, such as whether the subject takes a long or a short time to perform the task on any particular trial, may not be recorded at all.

The observational terms in both field and captive studies are operationalized, but in some cases operationalization differs between the two contexts. For example, take the case of cooperation in chimpanzees. Field researchers have reported that some chimpanzees spend years learning how to work together to hunt monkeys (Boesch 1994, 2002, 2005). Lab researchers have reported that chimpanzees fail to cooperate, because they do not have shared goals. For example, even when chimpanzees work together in order to access food, they do so in order to gain food for themselves, not for a shared sense of "us" (Tomasello 2016). Here it appears that the terms used to define "cooperate" differ in the field and lab studies. For the field, "cooperate" refers to working together, whereas in the lab (or at least some labs) "cooperate" requires having shared goals. Just as we need to keep a common language to compare humans with other animals, we need a common language to synthesize our knowledge of apes in the field and in the lab.

Evidence that chimpanzees lack shared goals, according to Tomasello, is that in a lab study when chimpanzees are given the chance to work alone or to work with another chimpanzee, they choose to work alone, whereas children who are given the chance to work alone or with other children tend to choose working with others (Rekers et al. 2011). Because children prefer to work together, it is presumed that they are motivated by what they perceive to be a shared goal. But because chimpanzees prefer to work alone, it is presumed that they are not motivated by the perception of a shared goal, but by the reward on offer.

There are compounding problems with this debate over the nature of cooperation. For one, since the researchers are using the same term differently, what appears to be a disagreement may not be. This problem is amplified when we use a cognitive mechanism in the definition and presume that all human instances of the behavior involve that cognitive mechanism. When humans work together, we may sometimes be motivated by shared goals, and sometimes by individual goals, but regardless we call it cooperation. Erdogan seeks cooperation with Germany, and Manafort cooperates with the Special Council, but the cooperative partners in these cases are unlikely to share the same goal.

The behavioral definition of "cooperation" at play in the field research is silent on mechanism; it follows common usage as applied to human behavior. Given that there are vast differences in kinds of human cooperation – from two children digging a hole together, to a criminal trading testimony for a reduced sentence, to scientific collaborators coauthoring a paper when they have never even met – a definition of "cooperation" in terms of mechanism is likely premature. Yet another problem arises with this case – how can we compare human children with nonhuman great apes?

Finally, at least some topics of investigation will be different for captive and wild apes. Field researchers can examine issues such as territoriality, war, immigration, communication, culture, social norms, and cooperation from an anthropological perspective right within the community they are studying. They can observe apes' behavior in order to describe whether they participate in these sorts of practices, and there is evidence that they do (Andrews 2020). Lab researchers can investigate more specific questions within these topics, and they can generate additional evidence in support of a field observation. Using experimental methods, they can control the experimental context to determine relevant variables leading to some observed behaviors. For example, field researchers can catalogue the gestures that great apes use to intentionally communicate (Hobaiter and Byrne 2011), while lab researchers can produce additional evidence that these gestures are intentionally produced (Tomasello et al. 1985). Field researchers can observe whether immigrants to new communities modify their tool use to conform (they at least sometimes do) (Luncz and Boesch 2014). Lab researchers can investigate whether apes prefer to imitate behaviors demonstrated to them or behaviors that they discover on their own (they tend to prefer the demonstrated behaviors) (Bonnie et al. 2007).

4.3.2 Biases in Field Research

The classic worry about bias in field research is based on the observational methods used in much of the research, which is what I will be focusing on here.

While some field researchers use experimental methods, for example, by recording alarm calls and doing playback experiments, or by placing rubber snakes on a path to observe who will give alarm calls, for the most part field research is "merely" observational. This owes largely to their scientific interests – how animals behave in their everyday lives – but also to ethical concerns.

Ape field research is typically conducted by people trained in anthropology, biology, or primatology, and the methods reflect those disciplinary backgrounds. Guides to field research begin by stressing the importance of learning how to see (Lehner 1996; Setchell and Curtis 2011). When a young student first goes to the field, it is difficult even to identify individuals. It takes time to learn the typical behavior of the species from directly observing them. In ethology, this preliminary period is called "reconnaissance observation." Observations are collected but not used as data for analysis. Rather, by collecting observations the new researcher comes to get to know the subjects and their living context and see both individual differences and species-normal practices – they develop knowledge of the usual. This baseline expectation can then be used to develop an ethogram and to engage in formal data collection.

Learning how to see is a skill, and as a skill it is also a source of bias. Those who are more or less skilled may end with different results. When behavior is videotaped and is being coded by multiple observers, disagreements in the interpretation of the behaviors are typically resolved through discussing which interpretation is most warranted, and that introduces another source of bias due to the relationships between the observers.

A common complaint about field research, especially when it comes to unusual behaviors, is that it amounts to mere anecdote – "The plural of 'anecdote' is not 'data.'" With anecdotes, we don't know how frequently such behaviors occur in the same sort of situation, making it possible that the behavior was an accident, like a cloud forming into the shape of a pumpkin. This appears to be Tomasello's complaint about the orangutan pantomime incidents. Without controls and repeated instances of the same movement in the same situation, he thinks we are not justified in taking Ceceb's behavior as intentional and meaningful. Furthermore, with anecdotes, we might have a naïve, egoistic description of behavior. The person might see just what they want to see, may be unfamiliar with the species, may be untutored in their practice of learning how to see, or be at an initial stage of their use of the principle of charity as a posit.

In response to the objection to the use of anecdotes, field researchers point out that they make "incident reports" (rather than using the vaguely insulting "anecdote") from a place of expertise. Just as humans don't tend to think that the clouds are trying to tell them something when they take the form of a pumpkin, field researchers have developed expertise in distinguishing meaningful from accidental behaviors. Like night nannies who specialize in infants and nurses who care for people with dementia, their regular interaction with their charges establishes an expertise and know-how that is not accessible to a naïve observer. This nondiscursively learnable proficiency, which I have called "folk expertise," makes the distinction between naïve or egoistic anecdote and a respectable incident report (Andrews 2009). A folk expert might be a PhD or a high school–educated field assistant. What makes them folk experts is their years of experience in the field.

For example, during Anne Russon's many years spent living in the forest with orangutans, following them from dawn to dark, with all the neck aches and bug bites that go along with it, she developed the skill of seeing orangutan behavior. Her descriptions of behavior allowed her to make predictions that turn out to be accurate, which in turn help to further justify those descriptions. She knows how to predict when an individual will be sweet and when an individual will be aggressive, or sneaky, or manipulative. She knows how to communicate with orangutans by understanding when they are asking for something, and when they want to be left alone. She knows how not to distract them, and where to find them in the forest. She knows how to anticipate where they might go next, and how far they have traveled given their activities and levels of agitation. Russon points out that her observations are informed ones, part of a systematic data collection by experienced observers, and are corroborated by other observations (Russon et al. 1993).

Incident reports can be part of a systematic study of animal behavior. Richard Byrne argues that "careful and unbiased recording of unanticipated or rare events, followed by collation and an attempt at systematic analysis, cannot be harmful. At worst, the exercise will be superseded and made redundant by methods that give greater control; at best, the collated data may become important to theory" (Byrne 1997, 135). One of the biggest research collaborative research projects in the 1980s relied on field researchers' incident reports. Byrne and Whiten's investigation into deception in primates was based on a large body of narratives describing candidate deceptive behaviors observed by field researchers (Whiten and Byrne 1988).

Depending on how a call for incident reports is made and who is reached by the call, such projects may introduce more bias than needed into the practice. A systematic way of collating incident reports could help to minimize bias and could be of even more value to theory and future research. Primatologists could develop a repository for observations that they tag with various descriptors, such as deception, pantomime, cooperation, punishment, norm violation, or teaching. When they observe an incident that serves as a candidate exemplar of one of these categories, they could record it in the repository, with any accompanying video. In this way, the observations could serve as a developing dataset, such as Wordbank, Stanford's open database of children's vocabulary development, or Avibase, a database recording the movements and behaviors of all the world's birds. With an open and well-publicized database, availability biases arising from the choice of who is asked for reports could be minimized.[3] Such a repository could also serve as an important window into primate natural living for those not in the field.

Though field researchers' observational methods have come under scrutiny, other sources of bias in field research have garnered much less attention. Field researchers well know that their mere presence changes the behavior of the animals they are studying, and even among habituated animals there may be observation effects. For example, orangutans in a new research site might react to human presence by running away, throwing branches at human observers, or kiss squeaking. When setting up a new field research site, it can take years to habituate apes to the presence of humans, and these are years in which good behavioral data is scarce. Even after habituation reduces these aggressive or fearful behaviors, human presence may continue to affect the animals. Orangutans may come down to the ground less often than they do when humans are not around, for example. Camera traps might be used to mitigate these effects, but for apes who spend a lot of time in trees or who travel vast distances camera traps are limited in the amount and kinds of behaviors they can capture. They tend to be more useful for studying food processing at known locations, such as at a termite mound, rather than for the study of behaviors such as mother–infant interaction, dominance battles, or nest building.

Another area that can create bias is in the researchers' choice of research area and camp location. For example, field camps in Borneo are often set up alongside the edges of a group's range, and nest-to-nest follows are usually limited by the camp's location. If an orangutan leaves the vicinity, researchers often stop following them. Furthermore, researchers rarely stay to observe the nighttime behavior of wild apes. Researchers have difficulty in travelling all the places that the apes travel. Given that humans lack apes' arboreality, we can't as easily cross deep gullies. We can't move though the dense forest as fast as apes can, and we often can't see what is happening high up in the trees. All of these limitations introduce systematic bias into field research.

While field research permits long-term and intergenerational anthropological, social, and cultural analyses, cognition may be difficult to study, since

---

[3] This idea has long been suggested by Anne Russon, but so far as I know it has not yet been taken up by any primatologist.

variables will be hard to manipulate, and rather than setting up situations to elicit behavior to be studied, researchers have to sit and wait for it. The human limitations of field research, and the choice of field and camp sites that are comfortable for researchers, may further impact what researchers can see. These are all sources of bias. However, the predominant criticism of fieldwork focusing on anecdote has been overblown.

4.3.3 Biases in Captive Research

Experimental captive research has the reputation of being the gold standard, especially where cognition is concerned. Psychologists can test their subjects in carefully controlled experiments, which can be repeated over time in the same lab and across labs with different researchers. However, there also exist many criticisms of captive research. In a systematic analysis of studies purporting to find humans to be uniquely intelligent, David Leavens and colleagues find that methodological, theoretical, and logical problems abound (Leavens et al.2017). The primatologist Christoph Boesch identifies five ways in which experimental studies on apes introduce variables that do not exist for human subjects:

a. Human subjects are selected from free-ranging individuals living in natural social groups, whereas ape subjects are selected from captive individuals living in deprived social groups;
b. Human subjects are tested with conspecifics, whereas ape subjects are tested with members of another species (normally humans);
c. Human subjects are tested in the same room as the experimenters, whereas ape subjects are separated by physical barriers from the experimenters;
d. Infant human subjects are in close proximity to one of their parents during testing, whereas infant ape subjects are separated from their biological mothers during testing;
e. Human subjects are tested about conspecific tasks with conspecific materials, whereas ape subjects are tested about human tasks with human materials.

Boesch (2007, 233–234)

Boesch concludes that these comparative studies cannot be used to claim that ape subjects lack a property that children have, because we are not putting the question to a fair test. In response to his critique, Tomasello and Call reply that these differences in studies are not problematic, because the issues of internal validity (variables b–e) are dealt with using control conditions, and the concern about external validity (variable a) is unwarranted (2008). It is worth examining this debate in some detail.

In response to the issues of internal validity, Tomasello and Call argue that control conditions suffice to minimize these worries. They argue that we know the variables such as not having a conspecific investigator or being given a species-typical task are extraneous, because apes are given control tasks to pass that have all these same variables attached. Since the apes can pass the control tasks, the variables are not relevant to control and test performance differences. For example, if apes are rewarded with food but children are rewarded with toys, researchers can be sure that the food reward is intrinsically motivating by testing it to motivate behavior in a control task.

In this response, Tomasello and Call are recognizing the biases of their studies and demonstrating how they attempt to deal with those biases. However, they don't consider that the variable may be relevant only for the experimental task and not the control task. A control task has to be of the right sort to serve as a true control, and the worry is that controls are not always designed to control for the relevant variable. A social variable may be more relevant in a social task, for example, than in an asocial task. Finding that the choice of researcher or the proximity to conspecifics didn't impact performance on a causal researching task doesn't allow us to infer that these features won't impact performance in a social task.

To illustrate, we can look at a debate between Tomasello and colleagues with Frans de Waal, Christophe Boesch, Victoria Horner, and Andrew Whiten (2008), in a response to a study finding that two-year-old children and apes have similar ability levels in physical technical tasks (with apes excelling in some of them), but that the children are superior to apes in social tasks (Herrmann et al. 2007).[4] De Waal and colleagues objected to the conclusions of the study, writing, "Human children sit on or next to their parent (creating potential 'Clever Hans' effects) and receive verbal instructions. They are used to dealing with strangers and are tested by a member of their own species. The apes are alone and confined, receive no verbal instructions, and are tested by a species not their own" (de Waal et al. 2008, 569). As we saw in Section 1, the flip side of Clever Hans effects is scaffolded practices that require the right kind of partner. When apes lack scaffolds and children have them, apes are also at a disadvantage.

Tomasello and Call agree that while a conspecific investigator or demonstrator would have been preferable in this study, such a setup wasn't feasible. They also suggest that the physical tasks, which used the same setup, served as

---

[4] Tomasello and Call note that in the Herrmann study the children were tested behind barriers in order to better make the testing situation parallel for the captive apes and the children.

a control condition. Because in the physical tasks the apes were sometimes better than the children, Tomasello and Call suggest that the differences in how the children and the apes were tested were controlled for and cannot be used to explain the results of the study.

However, this response neglects to consider why bias might emerge from using outgroup investigators and demonstrators. The physical tasks involved things such as spatial memory, object tracking, using noise cues to locate an object, discriminating quantities, and tool use. The social tasks involved things like social learning through observing a demonstrator, comprehending communicative cues to find an object, producing communicative cues to indicate the location of an object, gaze following to find an object, and understanding the intention behind a failed action. In both sets of conditions, the subjects had to learn how to solve a problem. In the physical tasks, there was no demonstrator. In the social tasks, there was a demonstrator – a communicative partner with information the subject could gain. The social tasks were largely tasks related to social learning. This is where the physical tasks and the social tasks come apart. In studies of human children, psychologists have come to realize the important role of selective social learning. Children are selective social learners – they won't learn from just anyone. They are less likely to imitate individuals who speak a different language (Buttlemann et al. 2013) or low-status individuals (Chudek et al. 2012).

Apes are also selective social learners (see Andrews forthcoming, Chapter 8 for a discussion). Apes prefer imitating high-ranking individuals (Bonnie et al. 2007). Rehabilitant orangutans prefer imitating some humans over others (Russon and Galdikas 1995). This is also true of monkeys. In an experiment conducted on wild vervet monkeys, monkeys who observed females retrieving food from a box via one of two doors tended to imitate the demonstrator, but monkeys who observed males performing the same behavior tended not to imitate (van de Waal et al. 2010). This makes sense, as vervet monkey males tend to immigrate, while females spend their lifespan in the same environment. The local experts are bound to be the females.

Given that apes engage in selective social learning, using out-group human demonstrators for social tasks should be expected to impact performance, while it should not be expected to impact performance on individual tasks that do not have the same social element. As was discussed in Section 1, failing to form relationships with an animal will augment this problem, as research participants will be more likely to take an unaffiliated individual as an out-group member.

To control for the bias of using in-group members to test children and out-group members to test chimpanzees, the children could be tested by people who speak a different language and who come from another cultural group – people who look, sound, and maybe even smell different from the children's typical group members. Another way to control for the bias is to test the animals using human experimenters who have a strong bond with the subject, making the human an in-group member. When apes are tested by volunteers or research assistants who are unfamiliar with apes in general, who don't have, or try to develop, relationships with these apes in particular, researchers are provoking the worst substitute teacher phenomenon. This is another reason why it is wrongheaded to teach students not to form relationships with their subjects, as the Anti-anthropomorphism principle suggests.

While Tomasello and Call acknowledge that there are biases in the ways the experiments are run, they think that they have controlled for all the variables.[5] However, a control is meant to isolate the independent variable that might explain the differences between the two groups, and the determination of what counts as a variable is where bias lies. The most obvious independent variable for explaining the difference in the physical and social tasks is that the children had appropriate social partners and the apes did not. Modifying the social partner is the only way to control for that variable.

The second response Tomasello and Call make to Boesch deals with variable (a), which they describe as a concern about external validity. Here they object that there is nothing wrong with captive apes as research subjects. Instead, Tomasello and Call suggest that there is some reason to think that captive apes might be more cognitively skilled than wild apes, because captive apes are given enrichment, which trains them to solve problems. Furthermore, because the apes are living with humans, they also have to learn to understand human behavior, communication, and norms.

However, to defend the idea that captive apes may be more sophisticated than wild apes one would have to review the cognitive demands that wild apes confront and captive apes do not, and to compare the respective demands. Wild apes have to learn to manage a host of skills and practices that are not present in captive apes, including finding and processing food, dealing with neighboring communities (including boundary patrols and territory incursions),

---

[5] More specifically, Tomasello and Call write, "in most cases we have control conditions for each species that have the same general task variables as the key experimental condition in terms of rewards, experimenters, housing situation, response requirements, and so forth – basically all of the variables Boesch identifies as problematic – and success with these control conditions is prerequisite for valid assessment in the experimental condition. In terms of species comparisons, it is only if both species pass these control conditions that the results comparing them in the experimental condition may be considered valid. Even so, recognizing the methodological differences, we mostly make our species comparisons not by statistically comparing the performance of the two species directly, but rather by statistically comparing the experimental and control conditions within each species separately and then comparing the pattern informally (e.g., one species is higher in experimental than control whereas the other is not)" (Tomasello and Call 2008, 449).

tracking irregular seasonality and fruiting production, the complexities of arboreal travel, keeping track of friendly and competitive conspecifics as they travel, social learning of traditional travel routes, social signals, hunting, how to select trees to climb and for building nests, how to care for infants, and so on.

In captivity, there is a relative lack of work. Captive apes don't need to learn how to process food or where to find water or when the trees will fruit. I would also expect less need to learn cultural practices. Captive apes will have lost many cultural practices due to lack of need to demonstrate them and lack of resources with which to enact them. For example, leaf-clipping to communicate sexual interest is a cultural behavior for chimpanzees in Mahale, but such a signal wouldn't be available to chimpanzees who are housed in enclosures without trees of the right sort or without possible sexual partners.

Another difference between the two populations is that apes in the wild have more autonomy. Wild apes get to decide when to wake, when to eat, where to travel, who to travel with, how far to go, where to nest, whether to immigrate to a new group, and so on. This kind of decisional authority does not exist for most captive apes, who are trained to live and sleep in one enclosure. Captive apes lack the freedom to leave the group and immigrate to a new one, which female chimpanzees often do when they reach adolescence. While this immigration process is sometimes emulated in zoo settings via animal transfers, the transfers are enforced and don't permit the individual to choose where, when, how, and whether to move.

We might also expect to see differences in the social relations, emotions, and flourishing between apes in the wild and apes in captivity. Good relationships between individuals is an important ingredient of a healthy life for social species, so in order to flourish captive apes need good relationships with the humans who care for them, the humans who test them, and the conspecifics they live with. It is difficult to manage captive animal groups. This is why caregivers at zoos and sanctuaries spend much time thinking about whether to move or separate individuals in order to provide a healthier social situation. Concern for the mental health of captive apes drives management decisions, though studies looking at the rates of mental disorders and abnormal behaviors in captive apes compared to wild populations find higher rates of mental problems among captive apes (Birkett and Newton-Fisher 2011; Ferdowsian et al. 2011, but see Rosati et al. 2013 for a critique and Ferdowsian et al. 2013 for their reply).

Tomasello and Call do not dispute the claim that Boesch makes about the different populations, but they do dispute his claim that captive chimpanzees are impaired as research subjects. I agree that it is unhelpful to speak of any animal as an impaired research subject; rather, the differences in individuals make them better suited for answering some research questions over others. That there are differences between wild and captive populations does raise questions about how well what we learn about captive chimpanzees generalizes to captive chimpanzees, and vice versa.

Boesch's claim about confounding variables in captive ape studies is not true of all captive research. However, keeping it in mind while designing experiments and interpreting results will help to minimize bias. These sources of bias are not exhaustive; we can add three more areas of bias to Boesch's list.

Bias may be introduced by researchers who lack folk expertise, the expertise developed through time spent in proximity and in relationship with the subject. For example, as we saw in Section 2, researchers tried and failed for almost forty years to elicit false belief tracking behavior, and it took thinking like a chimp to develop materials that held their interest. Without the folk expertise that led to the idea to test chimpanzees on a violent version of the false belief task, the community may have settled on the conclusion that apes aren't sensitive to others' false beliefs.

Field researchers often stress the importance of developing relationships with their subjects, whereas lab researchers are trained to avoid it. Just as Jane Goodall named chimpanzees and anthropologist Shirley Strum says that she wouldn't have understood olive baboons without living with them for so many years (Strum 2019), long-term field researchers tend to gain folk expertise the same way human caregivers do about infants. Folk expertise is the foundation for asking deeper or different questions. For example, the observation that some juvenile female chimpanzees carry sticks in particular ways led to the question of whether chimpanzees engage in something like pretend doll play (Kahlenberg and Wrangham 2010).

In addition to the biases that come from an enthusiasm about following the Anti-anthropomorphism principle, additional biases can arise when a researcher lacks expertise in the biology and ecology of the subject. A familiar critique of captive studies is that they lack ecological validity – that is, the tasks and setups don't reflect individuals' capacities in their natural environment. Worries about ecological validity arise when trying to generalize from the capacities of captive apes to the capacities of wild ones. They also arise when researchers use materials and experimental settings that differ significantly from the natural environment. For example, it wouldn't be ecologically valid to test cooperation in Western humans by asking them to share soup out of a single bowl. Likewise, given chimpanzees' tendency to fight over food, it wouldn't be ecologically valid to test cooperation in chimpanzees by asking them to share food.

While captive research permits careful control of variables, repeatability, and hence the ability to get at cognitive mechanisms, behaviors that require cultural transmission or knowledge as well as scaffolded cognitive abilities may be hard to uncover given the quality and longevity of the social group. The sources of bias that Boesch identified, along with those stemming from the quality of knowledge of the individual chimpanzees and species-typical wild behavior, and the quality of the relationships with captive subjects, are areas that may impact the outcome of the studies and should be ongoing topics of conversation among captive ape researchers.

4.4 Reconciling Field and Lab

No science is immune from bias, and all sciences have to work to reduce it. But without an explicit articulation of the potential sources of bias, it is difficult to know how to do so. Topics of investigation that are multidisciplinary should be expected to have more potential sources of bias, given the different methods, tools, and contexts in the different fields.

4.4.1 Best Practices for Multidisciplinarity

Any multidisciplinary endeavor will have to find ways of reaching past disciplinary boundaries. At minimum, multidisciplinarity requires that scientists agree roughly on the meaning of observational terms. This doesn't mean that all theoretical debates need to be resolved, but it does mean that participants have to agree on what the theoretical debates are *about*. When the meaning of a term such as "cooperation" is in dispute, it may be useful to revert back to the folk psychological sense of the term, coin a new term, or use a phrase to refer to the phenomenon rather than treat the term as if it reflects a shared perspective.

In adopting multidisciplinarity, we implicitly accept that having different types of evidence in favor of a hypothesis should make our credence in the hypothesis stronger than having only evidence of the same type. That is, if we have two reasons to think that apes are able to attribute false beliefs, but both reasons are the results of experimental studies on apes in moved object false belief tasks, then we have less reason to accept the conclusion than we would have with corroborating evidence from one experimental result and one field study in another context (Andrews 2018). There is virtue in having corroborating explanations of different types.

The goal of multidisciplinary collaborative research programs is to be able to defend an overhypothesis that best explains a rich set of interdisciplinary results. This approach speaks against the desirability of finding *the experiment* that would decide whether apes have some capacity. The dream of the one magic test characterizes the theory of mind research program, where the false belief task has been seen as *the* litmus test for chimpanzee false belief attribution. When apes failed versions of the test, researchers became skeptical that apes have mind-reading capacity; but once the apes passed two versions of the moved object false belief task, researchers took that to be sufficient evidence to conclude that apes understand false belief (Krupenye et al. 2016; Buttelmann et al. 2017). Now there is a third version, by the same team that published the first study (Kano et al. 2019).

An overhypothesis could be developed post hoc by doing a review of existing literature, but only if a rich enough body of literature were present. Preferably, an overhypothesis would emerge through a massive interdisciplinary team of researchers who know how to divide a question into parts that are best answered using the various expertises and methods present in the different disciplines.

Animal cognition research has a few examples of massive collaborative studies. The primate deception study by Whiten and Byrne introduced earlier is one example of this. In another study, using data collected at seven different field sites, researchers were able to determine that wild chimpanzee communities are culturally distinct, with some showing patterns of behavior that are absent in others (Whiten et al. 1999). However, this study focused only on data from the field. In another study, experimentalists examined the performance of thirty-six mammals and birds on two tasks meant to measure self-control in order to determine variables that best predicted strong performance (MacLean et al. 2014). The focus of this study was performance on the tasks and the neurological data, though information about dietary breadth and group size from field researchers also was examined. They found that absolute brain size best predicted performance on the two tasks, so while the field data was examined to see if it was relevant, it wasn't used to help motivate or design the study. Furthermore, researchers on avian species objected that the range of species wasn't representative and that corvids perform like apes (Kabadayi et al. 2016). Like with calls for incident reports from the field, calls for experimental research on different species could use some institutional support to make sure it is inclusive.

Massive collaborative studies that involve both field researchers and experimentalists could combine the best of our knowledge from both sources. By agreeing on how to define terms and by defining them cognitively neutrally to begin with, scientists will better be able to integrate our knowledge to develop a better understanding of great apes – and humans. Such collaborations, if they were to take off, would likely drive additional multidisciplinary methods and theories, which in turn might be sources for additional biases but would serve to promote the science of ape cognition.

4.4.2 When the Lab Is the Field

In ape research, the field is the field and the lab is the lab. While some experiments can be done in the field, the contexts remain quite different. This isn't true for all the species we study. Dog cognition – or dognition – research is more like human research. Humans and dogs coevolved in the same evolutionary context. Our world is their world. When we study dogs in a lab, dogs go with their human, play games, get treats and positive social reinforcement, then go home. When we study babies in a lab, the infants go with their caregiver, play games, get treats and positive social reinforcement, then go home.

Humans have also long shared territories with other primates, just not in North America or Europe, where much primate research is institutionally and financially supported. The fact that Japanese primatologists first identified cultures in the native macaque populations should come as little surprise, since the primatologists' antecedent proximity to their subjects served as a foundation for developing the relevant folk expertise.

Primates play a role in traditional human stories, and in many places primates and humans live side by side. Balinese macaques and human worshippers share temple grounds and have developed agreements about how to live together in harmony (Fuentes 2010). Monkeys are sensitive to which temple offerings they are allowed to eat, and which offerings they must not touch. They treat tourists as easy marks to steal from but leave the locals alone. Humans who guard the temple will intervene after a monkey steals a camera or sunglasses from a hapless tourist, offering one treat after another until the monkey decides the trade is worth it (Brotcorne et al. 2017). Orangutans in Borneo have lived with Dyak people in uneasy relations for hundreds if not thousands of years. Current conflicts between orangutans and humans in locals' fruit gardens, palm plantations, and mining operations are other places to examine ape behavior in a shared context.

Liminal animals who share the wild with humans, such as the raccoons, rats, and prairie dogs in our towns and cities, monkeys in our temples, and orangutans in our gardens, offer places to reconcile field and lab studies. Such in-between animals can be studied using the best of our methods from the field and from the lab. When we seek to understand the animals who live with us, we introduce some biases, but we also introduce some expertise. There will always be bias, wherever we look. But there is not always expertise. Risking bias is worth it in those contexts where we already have developed expertise – in our homes, towns, and gardens.

Conclusion

The principles of comparative psychology, including Anti-anthropomorphism, Morgan's Canon, and rules to avoid forming relationships with animals and not to presume anything about animal consciousness, have been introduced to minimize bias in the science. Rather than seeing animals as sentient beings who live in community and have their own interests, scientists are instructed by these principles to remain distant and detached. This means that the principles end up introducing different biases. The comparative psychologist John Gluck, who completed his PhD under Harry Harlow, describes his change of perspective regarding animals during his training as a comparative psychologist in the 1960s:

> [T]hose attracted to a career involving research on animals must undergo an emotional and ethical retraining process every bit as important as their scientific training . . .
>
> I grew up with deep emotional attachments to family pets, believed without question that animals had internal lives that mattered to them and were capable of feeling joy, sadness, fear, disappointment, and pain, and was revolted by cruelty to animals . . . By the time I had finished my undergraduate education and started graduate school, my professors – and the overall research context into which I threw myself – had exorcised my sentimental concern for animals' welfare and constructed for me a new belief system in which there was really no such thing as the animals' perspective. (Gluck 2016, 13–14)

Gluck was taught not to take his research participants as sentient beings with interests of their own and with value, all in the name of science, but as machines from which to extract scientific knowledge valuable for humans. This attitude impacts the quality of the science, because it leaves topics, perspectives, and hypotheses unexamined, and it provides a sterile research context in which to ask questions. Furthermore, such attitudes can also impact the morality of the research. In his book, Gluck attempts to explain how his training led him to conduct maternal deprivation and social isolation research on primates that he now sees as immoral. Given the training, he couldn't see the ethical problems as they arose.

The science of comparative psychology can best proceed by rejecting a mechanistic view of animals and instead seeing them as sentient beings who live in community, even as they see them as individuals who are in many ways very different from us. Scientists who study humans treat them as sentient beings with whom they can have relationships while at the same time recognizing that their perspectives may differ greatly from the scientists' own, and that the relationships they have with their participants may lead to bias. Comparative psychologists should be trusted to handle the same sources of bias, especially given the ethical costs of refusing to do so.

Bibliography

Allen, C. (2004). Is anyone a cognitive ethologist? *Biology and Philosophy*, *19*(4), 589–607.

Allen, C., & Bekoff, M. (1997). *Species of Mind: The Philosophy and Biology of Cognitive Ethology*. Cambridge, MA: Massachusetts Institute of Technology Press.

Allen, C., Fuchs, P. N., Shriver, A., & Wilson, H. D. (2005). Deciphering animal pain. In Aydede, M. (Ed.), *Pain: New Essays on Its Nature and the Methodology of Its Study*, 351–366. Cambridge, MA: Massachusetts Institute of Technology Press.

Allen, C., & Trestman, M. (2017). Animal consciousness. In Zalta, E. N. (Ed.), *The Stanford Encyclopedia of Philosophy (Winter 2017 Edition)*. https://plato .stanford.edu/archives/win2017/entries/consciousness-animal

Allen, J. A. (2019). Community through culture: From insects to whales. *BioEssays*, *41*(11), 1900060.

Andrews, K. (2007). Critter psychology: On the possibility of nonhuman animal folk psychology. In Hutto, D., & Ratcliffe, M. (Eds.), *Folk Psychology Reassessed*, 191–210. New York, NY: Springer.

Andrews, K. (2009). Politics or metaphysics? On attributing psychological properties to animals. *Biology & Philosophy*, *24*(1), 51–63.

Andrews, K. (2011) Beyond anthropomorphism: Attributing psychological properties to animals. In Beauchamp, T., & Frey, R. G. (Eds.), *The Oxford Handbook of Animal Ethics*, 469–494. Oxford, NY: Oxford University Press.

Andrews, K. (2012). *Do Apes Read Minds? Toward a New Folk Psychology*. Cambridge, MA: Massachusetts Institute of Technology Press.

Andrews, K. (2015). The folk psychological spiral: Explanation, regulation, and language. *The Southern Journal of Philosophy*, *53* (S1), 50–67.

Andrews, K. (2016a). Snipping or editing? Parsimony in the chimpanzee mind-reading debate. *Metascience*, *25*(3), 377–386.

Andrews, K. (2016b). A role for folk psychology in animal cognition research. In Blank, A. (Ed.), *Animals: Basic Philosophical Concepts*, 205–226. Munich, Germany: Philosophia.

Andrews, K. (2016c). Pluralistic folk psychology in humans and other apes. In Kiverstein, J. (Ed.), *The Routledge Handbook of Philosophy of the Social Mind*, 133–154. London, UK: Routledge.

Andrews, K. (2018). Do apes attribute beliefs to predict behavior? A Mengzian social intelligence hypothesis. *The Harvard Review of Philosophy.* doi:10.5840/harvardreview201892117.

Andrews, K. (2020). Naïve normativity: The social foundation of moral cognition. *Journal of the American Philosophical Association, 6*(1), 36–56. doi:10.1017/apa.2019.30

Andrews, K. (forthcoming). *The Animal Mind: An Introduction to the Philosophy of Animal Cognition*, second edition. Abingdon: Routledge.

Andrews, K., & Huss, B. (2014). Anthropomorphism, anthropectomy, and the null hypothesis. *Biology & Philosophy, 29*(5), 711–729.

Asquith, P. J. (1996) Japanese science and western hegemonies. Primatology and the limits set to questions. In Nader L. (Ed.), *Naked Science. Anthropological Inquiry into Boundaries, Power, and Knowledge*, 239–256. New York: Routledge.

Avital, E., & Jablonka, E. (Eds.) (2000). *Animal Traditions: Behavioural Inheritance in Evolution*. Cambridge, MA: Cambridge University Press.

Balter, M. (2012). 'Killjoys' challenge claims of clever animals. *Science, 335*(6072), 1036–1037.

Barrett, L. (2011). Beyond the Brain: How Body and Environment Shape Animal and Human Minds. Princeton, NJ: Princeton University Press.

Barron, A. B., & Klein, C. (2016). What insects can tell us about the origins of consciousness. *Proceedings of the National Academy of Sciences, 113*(18), 4900–4908.

Bekoff, M. (2007). *The Emotional Lives of Animals: A Leading Scientist Explores Animal Joy, Sorrow, and Empathy – and Why They Matter.* Novato, CA: New World Library.

Berridge, K. C., & Kringelbach, M. L. (2015). Pleasure systems in the brain. *Neuron, 86*(3), 646–664.

Bickle, J. (1998). *Psychoneural Reduction: The New Wave.* Cambridge, MA: Massachusetts Institute of Technology Press.

Birch, J. (2017). Animal sentience and the precautionary principle. *Animal Sentience, 2*(16), 1.

Birkett, L. P., & Newton-Fisher, N. E. (2011). How abnormal is the behaviour of captive, zoo-living chimpanzees? *PloS One, 6*(6), e20101.

Boesch, C. (1994). Cooperative hunting in wild chimpanzees. *Animal Behavior, 48*, 653–667.

Boesch, C. (2002). Cooperative hunting roles among Taï chimpanzees. *Human Nature, 13*, 27–46.

Boesch, C. (2005). Joint cooperative hunting among wild chimpanzees: Taking natural observations seriously. *Behavioral and Brain Sciences, 28*, 692.

Boesch, C. (2007). What makes us human (Homo sapiens)? The challenge of cognitive cross-species comparison. *Journal of Comparative Psychology*, *121*, 227–240.

Boffey, P. M. (1987, November 24). Infants' sense of pain is recognized, finally. *The New York Times*. www.nytimes.com/1987/11/24/science/infants-sense-of-pain-is-recognized-finally.html

Boisseau, R. P., Vogel, D., & Dussutour, A. (2016). Habituation in non-neural organisms: evidence from slime moulds. *Proceedings of the Royal Society B*, *283*(1829), 20160446. doi:10.1098/rspb.2016.0446.

Bonnie, K. E., Horner, V., Whiten, A., & de Waal, F. B. M. (2007). Spread of arbitrary conventions among chimpanzees: A controlled experiment. *Proceedings of the Royal Society B*, *274*, 367–372.

Brosnan, S. F., Schiff, H. C., & de Waal, F. B. M. (2005). Tolerance for inequity may increase with social closeness in chimpanzees. *Proceedings of the Royal Society B: Biological Sciences*, *272*(1560), 253–258.

Brotcorne, F., Giraud, G., Gunst, N., Fuentes, A., Wandia, I. N., Beudels-Jamar, R. C., ... Leca, J. B. (2017). Intergroup variation in robbing and bartering by long-tailed macaques at Uluwatu Temple (Bali, Indonesia). *Primates*, *58*(4), 505–516.

Brown, C. (2015). Fish intelligence, sentience and ethics. *Animal Cognition*, *18*(1), 1–17. doi:10.1007/s10071-014-0761-0.

Buckner, C. (2013). Morgan's Canon, meet Hume's Dictum: Avoiding anthropofabulation in cross-species comparisons. *Biology & Philosophy*, *28*(5), 853–871.

Burghardt, G. M. (1991). Cognitive ethology and critical anthropomorphism: A snake with two heads and hognose snakes that play dead. In C. A. Ristau (Ed.), *Cognitive Ethology: The Minds of Other Animals: Essays in Honor of Donald R. Griffin* (pp. 53–90). Hillsdale, NJ: Lawrence Erlbaum Associates.

Burghardt, G. M. (2005). *The Genesis of Animal Play*. Cambridge, MA: Massachusetts Institute of Technology Press.

Burghardt, G. M. (2007). Critical anthropomorphism, uncritical anthropocentrism, and naive nominalism. *Comparative Cognition & Behavior Reviews*, *2*, 136–138.

Burkhardt, R. W. (2005). *Patterns of Behavior: Konrad Lorenz, Niko Tinbergen, and the Founding of Ethology*. Chicago, IL: University of Chicago Press.

Buttelmann, D., Buttelmann, F., Carpenter, M., Call, J., & Tomasello, M. (2017). Great apes distinguish true from false beliefs in an interactive helping task. *PLOS ONE*, *12*(4), e0173793.

Buttelmann, D., Zmyj, N., Daum, M., & Carpenter, M. (2013). Selective imitation of in-group over out-group members in 14-month-old infants. *Child Development, 84*(2), 422–428.

Byrne, R. (1997). What's the use of anecdotes? Distinguishing psychological mechanisms in primate tactical deception. In Mitchell, R., Thompson, N., & Miles, H. L. (Eds.), *Anthropomorphism, Anecdotes, and Animals*, 134–150. Albany, NY: State University of New York Press.

Call, J., & Tomasello, M. (1999). A nonverbal false belief task: The performance of children and great apes. *Child Development, 70*, 381–395.

Calvo, P. (2016). The Philosophy of plant neurobiology: a manifesto. *Synthese, 193*(5), 1323–1343. https://doi.org/10.1007/s11229-016-1040-1

Carruthers, P. (2004). Suffering without subjectivity. *Philosophical Studies, 121*, 99–125.

Cheney, D. L., & Seyfarth, R. M. (1990). *How Monkeys See the World: Inside the Mind of Another Species*. Chicago, IL: University of Chicago Press.

Chollet, Francois. (2019). On the measure of intelligence. arXiv:1911.01547

Chudek, M., Heller, S., Biro, S., & Henrich, J. (2012). Prestige-biased cultural learning: Bystander's differential attention to potential models influencing children's learning. *Evolution and Human Behavior, 33*, 46–54.

Churchland, P. M. (1981). Eliminative materialism and the propositional attitudes. *The Journal of Philosophy, 78*, 67–90.

Clay, Z., & Tennie, C. (2018). Is overimitation a uniquely human phenomenon? Insights from human children as compared to bonobos. *Child Development, 89*(5), 1535–1544.

Clayton, N., & Dickinson, A. (1998). Episodic-like memory during cache recovery by scrub jays. *Nature 395*, 272–274. https://doi.org/10.1038/26216

Coulacoglou, C., & Saklofske, D. H. 2017. Perspectives and advances in personality, in *Psychometrics and Psychological Assessment*, 223–265. Cambridge, MA: Academic Press.

Crick, F., & Koch, C. (1990). Towards a neurobiological theory of consciousness. *Seminars in the Neurosciences, 2*, 263–275.

Dacey, M. 2017. A new view of association and associative models. In Andrews, K., & Beck, J. (Eds.), *The Routledge Handbook of Philosophy of Animal Minds*, 419–426. New York: Routledge.

Darwin, C. (1871). *The Descent of Man*. Mineola, NY: Dover.

Dawkins, M. S. (2008). The science of animal suffering. *Ethology, 114*(10), 937–945. doi:10.1111/j.1439-0310.2008.01557.x.

Dawkins, M. S. (2017). Animal welfare with and without consciousness. *Journal of Zoology, 301*(1), 1–10. doi:10.1111/jzo.12434.

Dennett, D. C. (1983). Intentional systems in cognitive ethology: The "Panglossian Paradigm" defended. *Behavioral and Brian Sciences, 6*, 343–390.

Dennett, D. C. (1991). Real patterns. *The Journal of Philosophy, 88*(1), 27–51.

Ferdowsian, H. R., Durham, D. L., Kimwele, C., Kranendonk, G., Otali, E., Akugizibwe, T., Mulcahy, J. B., Ajarova, L., and Johnson, C. M. (2011). Signs of mood and anxiety disorders in chimpanzees. *PLoS ONE, 6*(6), Article e19855.

Ferdowsian, H., Durham, D., & Brüne, M. (2013). Mood and anxiety disorders in chimpanzees (pan troglodytes): A response to Rosati et al. (2012). *Journal of Comparative Psychology, 127*(3), 337–340. https://doi.org/10.1037/a0032823

Fitzpatrick, S. (2008). Doing away with Morgan's Canon. *Mind & Language, 23*(2), 224–246.

Fitzpatrick, S. (2009). The primate mindreading controversy: A case study in simplicity and methodology in animal psychology. In R. W. Lurz (Ed.), *The Philosophy of Animal Minds* (pp. 224–246). Cambridge University Press.

Fitzpatrick, S. (2017). Against Morgan's Canon. In Andrews, K., & Beck, J. (Eds.), *Routledge Handbook of Philosophy of Animal Minds*, 437–447. New York, NY: Routledge.

Franks, B., Sebo, J., & Horowitz, A. (2018). Fish are smart and feel pain: What about joy? *Animal Sentience, 3*(21), 16.

Fuentes, A. (2010) Naturecultural encounters in Bali: Monkeys, temples, tourists, and ethnoprimatology. *Cultural Anthropology 25*(4), 600–624.

Gagliano, M., Renton, M., Depczynski, M., & Mancuso, S. (2014). Experience teaches plants to learn faster and forget slower in environments where it matters. *Oecologia, 175*(1), 63–72. https://doi.org/10.1007/s00442-013-2873-7

Gagliano, M., Vyazovskiy, V. V., Borbély, A. A., Grimonprez, M., & Depczynski, M. (2016). Learning by association in plants. *Scientific Reports, 6*, 38427. doi:10.1038/srep38427.

Gergely, G., Bekkering, H., & Király, I. (2002). Developmental psychology: Rational imitation in preverbal infants. *Nature, 415*(6873), 755.

Giere, R. N. (1988). *Explaining Science: A Cognitive Approach.* Chicago, IL: University of Chicago Press.

Giere, R. N. (2006). *Scientific Perspectivism.* Chicago, IL: University of Chicago Press.

Gluck, J. P. (2016). *Voracious Science and Vulnerable Animals: A Primate Scientist's Ethical Journey.* Chicago, IL: University of Chicago Press.

Goodenough, W. H. (1956) Residence rules. *Southwest Journal of Anthropology 12*(1), 22–37.

Griffin, D. R. (1976). *The Question of Animal Awareness: Evolutionary Continuity of Mental Experience.* New York, NY: Rockefeller University Press.

Griffith, C. R. (1943). *Principles of Systematic Psychology.* Champaign, IL: University of Illinois Press.

Harriman, P. L. (1947). *The New Dictionary of Psychology.* New York, NY: Philosophical Library.

Healy, S. (Ed.). (1998). *Spatial Representation in Animals.* New York, NY: Oxford University Press.

Herrmann, E., Call, J., Hernàndez-Lloreda, M. V., Hare, B., & Tomasello, M. (2007). Humans have evolved specialized skills of social cognition: The cultural intelligence hypothesis. *Science, 317*(5843), 1360–1366.

Heyes, C. (2008). Beast machines? Questions of animal consciousness. In Weiskrantz, L., & Davies, M. (Eds.), *Frontiers of Consciousness*, 259–274. Oxford, NY: Oxford University Press.

Hobaiter, C., & Byrne, R. W. (2011). The gestural repertoire of the wild chimpanzee. *Animal Cognition, 14*(5), 745–767.

Hodos, W., & Campbell, C. B. G. (1969). Scala naturae: Why there is no theory in comparative psychology. *Psychological Review, 76*(4), 337–350.

Hoppitt, W., & Laland, K. N. (2013). *Social Learning: An Introduction to Mechanisms, Methods,and Models.* Princeton: Princeton University Press.

Horner, V., & Whiten, A. (2005). Causal knowledge and imitation/emulation switching in chimpanzees (*Pan troglodytes*) and children (*Homo sapiens*). *Animal Cognition, 8*(3), 164–181.

Huber, L., Popovová, N., Riener, S., Salobir, K., & Cimarelli, G. (2018). Would dogs copy irrelevant actions from their human caregiver? *Learning & Behavior, 46*(4), 387–397.

Huber et al. forthcoming. Selective social learning in canids. *Learning and Behavior.*

Hutto, D. D., & Myin, E. (2012). *Radicalizing Enactivism: Basic Minds without Content.* Cambridge, MA: Massachusetts Institute of Technology Press.

Imanishi, K. (1957). Social behavior in Japanese monkeys, Macaca Fuscata. *Psychologia*, 1, 47.

Jensen, K., Silk, J., Andrews, K., Bshary, R., Cheney, D., Emery, N., … Teufel, C. (2011). Social knowledge. In Menzel, R. and Fischer, J. (Eds.), *Animal Thinking: Contemporary Issues in Comparative Cognition*, 267–291. Cambridge, MA: Massachusetts Institute of Technology Press.

Johnson, L. S. M., & Lazaridis, C. (2018). The sources of uncertainty in disorders of consciousness. *AJOB Neuroscience, 9*(2), 76–82.

Kabadayi, C., Taylor, L. A., von Bayern, A. M. P., & Osvath, M. (2016). Ravens, New Caledonian crows and jackdaws parallel great apes in motor self-regulation despite smaller brains. *Royal Society Open Science, 3*(4).

Kahlenberg, S. M., & Wrangham, R. W. (2010). Sex differences in chimpanzees' use of sticks as play objects resemble those of children. *Current Biology: CB, 20*(24), R1067–1068.

Kahsai, L., & Zars, T. (2011). Learning and memory in *Drosophila*: Behavior, genetics, and neural systems. *International Review of Neurobiology, 99*, 139–167.

Kano, F., Krupenye, C., Hirata, S., Tomonaga, M., & Call, J. (2019). Great apes use self-experience to anticipate an agent's action in a false-belief test. *Proceedings of the National Academy of Sciences, 116*(42), 20904–20909.

Keeley, B. L. (2004). Anthropomorphism, primatomorphism, mammalomorphism: Understanding cross-species comparisons. *Philosophy and Biology, 19*, 521–540.

Krupenye, C., Kano, F., Hirata, S., Call, J., & Tomasello, M. (2016). Great apes anticipate that other individuals will act according to false beliefs. *Science, 354*(6308), 110–114.

Kuhn, T. (1962). *The Structure of Scientific Revolutions*. Chicago, IL: University of Chicago Press.

Kummer, H. (1995). *In Quest of the Sacred Baboon*. Princeton, NJ: Princeton University Press.

Laland, K. N., & Janik, V. M. (2006). The animal cultures debate. *Trends in Ecology and Evolution, 21*(10), 542–547.

Lapuschkin, S., Wäldchen, S., Binder, A., Montavon, G., Samek, W., & Müller, K.-R. (2019). Unmasking Clever Hans predictors and assessing what machines really learn. *Nature Communications, 10*(1), 1–8.

Leavens, D. A., Bard, K. A., & Hopkins, W. D. (2019). The mismeasure of ape social cognition. *Animal Cognition, 22*(4), 487–504. https://doi.org/10.1007/s10071-017-1119-1

Lehner, P. N. (1996). *Handbook of Ethological Methods*. Cambridge: Cambridge University Press.

Lewis, D. (1972). Psychophysical and theoretical identifications. *Australasian Journal of Philosophy, 50*(3), 249–258.

Longino, H., 1990. *Science as Social Knowledge: Values and Objectivity in Scientific Inquiry*. Princeton: Princeton University Press.

Luncz, L. V., & Boesch, C. (2014). Tradition over trend: Neighboring chimpanzee communities maintain differences in cultural behavior despite frequent immigration of adult females. *American Journal of Primatology, 76*(7), 649–657.

Lyons, D. E., Young, A. G., & Keil, F. C. (2007). The hidden structure of overimitation. *Proceedings of the National Academy of Sciences, 104*(50), 19751–19756.

MacLean, E. L., Hare, B., Nunn, C. L., Addessi, E., Amici, F., Anderson, R. C., ... Zhao, Y. (2014). The evolution of self-control. Proceedings of the National Academy of Sciences, *111*(20), E2140–E2148. https://doi.org/ 10.1073/pnas.1323533111

Malik, B. R., & Hodge, J. J. L. (2014). Drosophila adult olfactory shock learning. *Journal of Visualized Experiments : JoVE*, (90).

Mather, J. A., & Dickel, L. (2017). Cephalopod complex cognition. *Current Opinion in Behavioral Sciences*, *16*, 131–137.

Matsuzawa, T. 2006. Sociocognitive development in chimpanzees: A synthesis of laboratory work and fieldwork. In Matsuzawa, T., Tomonaga, M., & Tanaka, M. (Eds.). *Cognitive Development in Chimpanzees*. Springer Japan. https://doi.org/10.1007/4-431-30248-4

Matsuzawa, T. (2007). Comparative cognitive development. *Developmental Science 10*(1), 97–103.

McLaren, I. P. L., & Mackintosh, N. J. (2000). An elemental model of associative learning: I. latent inhibition and perceptual learning. *Animal Learning & Behavior*, *28*(3), 211–246.

Meltzoff, A. N. (1988). Infant imitation after a 1-week delay: Long-term memory for novel acts and multiple stimuli. *Developmental Psychology*, *24*(4), 470–476.

Midgley, M. (2001). Being objective. *Nature*, *410*(6830), 753.

Mikhalevich, I. (2017). Simplicity and cognitive models: Avoiding old mistakes in new experimental contexts. In Andrews, K., & Beck, J. (Eds.), *The Routledge Handbook of Philosophy of Animal Minds*. New York: Routledge.

Milius, S. (2010, Aug. 10). Orangutans can mime their desires. *Science News Web Edition*. www.sciencenews.org/view/generic/id/61991/title/ Orangutans_can_mime_their_desires.

Morgan, C. L. (1891). *Animal life and intelligence*. London, UK: E. Arnold. http://archive.org/details/animallifeintell00morguoft

Morgan C. L. (1903). *Introduction to Comparative Psychology*. London, UK: Walter Scott.

Morgan, C. L. (1930). Autobiography of C. Lloyd Morgan. In Murchison, C. (Ed.), *History of Psychology in Autobiography*, 237–264. Worcester, MA: Clark University Press.

Myowa-Yamakoshi, M., & Matsuzawa, T. (2000). Imitation of intentional manipulatory actions in chimpanzees (*Pan troglodytes*). *Journal of Comparative Psychology*, *114*(4), 381.

Nagel, T. (1979). What is it like to be a bat? In *Mortal Questions*. Cambridge: Cambridge University Press.

Nisbett, R. E., & Wilson, T. D. (1977). Telling more than we can know: Verbal reports on mental processes. *Psychological Review*, *84*, 231–259.

Öhman, A., & Mineka, S. (2001). Fears, phobias, and preparedness: Toward an evolved module of fear and fear learning. *Psychological Review, 108*(3), 483.

Olds J. (1956, October). Pleasure centers in the brain. *Scientific American, 195*(4), 105–117.

Olds, J., & Milner, P. (1954). Positive reinforcement produced by electrical stimulation of septal area and other regions of rat brain. *Journal of Comparative and Physiological Psychology, 47*(6), 419–427.

Panksepp, J. (2016). Brain processes for "good" and "bad" feelings: How far back in evolution? *Animal Sentience, 3*(24). https://animalstudiesrepository .org/animsent/vol1/iss3/24/

Panksepp, J., & Burgdorf, J. (2003). "Laughing" rats and the evolutionary antecedents of human joy? *Physiology & Behavior, 79*(3), 533–547.

Panoz-Brown, D., Iyer, V., Carey, L. M., Sluka, C. M., Rajic, G., Kestenman, J., ... Crystal, J. D. (2018). Replay of episodic memories in the rat. *Current Biology, 28*(10), 1628–1634.e7. https://doi.org/10.1016/j.cub.2018.04.006

Pearce, J. M. (2008). *Animal Learning and Cognition, 3rd Edition*. Hove, NY: Psychology Press.

Penn, D. C. (2011). How folk psychology ruined comparative psychology: And how scrub jays can save it. In Menzel, R., & Fischer, J. (Eds.), *Animal Thinking: Contemporary Issues in Comparative Cognition*, 253–266. Cambridge, MA: Massachusetts Institute of Technology Press.

Penn, D. C., & Povinelli, D. J. (2007). On the lack of evidence that non-human animals possess anything remotely resembling a "theory of mind." *Philosophical Transactions of the Royal Society B, 362*, 731–744.

Povinelli, D. J., & Bering, J. M. (2002). The mentality of apes revisited. *Current Directions in Psychological Science, 11*, 115–119.

Povinelli, D. J. (2004). Behind the ape's appearance: Escaping anthropocentrism in the study of other minds. *Daedalus, 133*(1), 29–41.

Premack, D., & Woodruff, G. (1978). Does the chimpanzee have a theory of mind? *Behavioral and Brain Sciences, 1*, 515–526.

Proctor, H. S., Carder, G., & Cornish, A. R. (2013). Searching for animal sentience: A systematic review of the scientific literature. *Animals, 3*(3), 882–906. doi:10.3390/ani3030882

Ramsey, G. (2017). What is animal culture? In Andrews, K., & Beck, J. (Eds.), *The Routledge Handbook of Philosophy of Animal Minds*, 345–353. New York, NY: Routledge.

Rekers, Y., Haun, D. B. M., & Tomasello, M. (2011). Children, but not chimpanzees, prefer to collaborate. *Current Biology, 21*(20), 1756–1758.

Rendell, L., & Whitehead, H. (2001). Culture in whales and dolphins. *Behavioural and Brain Sciences, 24*, 309–382.

Robbins, J., & Rumsey, A. (2008). Introduction: Cultural and linguistic anthropology and the opacity of other minds. *Anthropological Quarterly, 81*(2), 407–420.

Rosati, A. G., Herrmann, E., Kaminski, J., Krupenye, C., Melis, A. P., Schroepfer, K., Tan, J., Warneken, F., Wobber, V., and Hare, B. (2013). Assessing the psychological health of captive and wild apes: A response to Ferdowsian et al. (2011). *Journal of Comparative Psychology, 127*(3), 329–336.

Russon, A. E. (2018). Pantomime and imitation in great apes: Implications for reconstructing the evolution of language. *Interaction Studies, 19*(1–2), 200–215. https://doi.org/10.1075/is.17028.rus

Russon, A., & Andrews, K. (2010). Orangutan pantomine: Elaborating the message. *Biology Letters, 7*(4), 627–630. https://doi.org/10.1098/rsbl.2010.0564

Russon, A., & Galdikas, B. (1993). Imitation in free-ranging rehabilitant orangutans. *Journal of Comparative Psychology, 107*, 147–161.

Salas, C., Broglio, C., Dúran, E., Gómez, A., Ocana, F. M., Jimenez-Moya, F., & Rodriguez, F. (2006). Neuropsychology of learning and memory in teleost fish. *Zebrafish, 3*, 157–171.

Schwitzgebel, Eric. 2016. Phenomenal consciousness, defined and defended as innocently as I can manage. *Journal of Consciousness Studies, 23*, 11–12, 224–235.

Setchell, J. M., & Curtis, D. J. (2011). *Field and Laboratory Methods in Primatology: A Practical Guide* (second edition). Cambridge, UK: Cambridge University Press.

Seyfarth, R. M., Cheney, D. L., & Marler, P. (1980). Vervet monkey alarm calls: Semantic communication in a free-ranging primate. *Animal Behavior, 28*, 1070–1094.

Shettleworth, S. J. (1993). Where is the comparison in comparative cognition?: Alternative research programs. *Psychological Science, 4*(3), 179–184.

Shettleworth, S. J. (2010a). Clever animals and killjoy explanations in comparative psychology. *Trends in Cognitive Science, 14*, 477–481.

Shettleworth, S. J. (2010b). *Cognition, Evolution, and Behavior* (2nd Edition). New York, NY: Oxford University Press.

Shettleworth, S. J. (2012). *Fundamentals of Comparative Cognition*. Oxford: Oxford University Press.

Shriver A. (2018). The unpleasantness of pain for humans and other animals. In Bain, D., Brady, M., & Corns, J. (Eds.), *Philosophy of Pain: Unpleasantness, Emotion, and Deviance*, 147–162. New York, NY: Routledge.

Silk, J. B. (2002). Using the 'F'-word in primatology. *Behaviour*, *139*(2–3), 421–446.

Sneddon, L. U. (2015). Pain in aquatic animals. *Journal of Experimental Biology*, *218*(7), 967–976. https://doi.org/10.1242/jeb.088823

Sober, E. (1998). Morgan's Canon. In Cummins, D., & Allen, C. (Eds.), *The Evolution of Mind*, 224–242. New York, NY: Oxford University Press.

Sober, E. (2005) Comparative psychology meets evolutionary biology: Morgan's Canon and cladistic parsimony. In Daston, L., & Mitman, G. (Eds.), *Thinking with Animals: New Perspectives on Anthropomorphism*, 85–99. New York, NY: Columbia University Press.

Sober, E. (2015). *Ockham's Razors: A User's Manual*. Cambridge, MA: Cambridge University Press.

Song, H., Fang, F., Tomasson, G., Arnberg, F.K., Mataix-Cols, D., Cruz, L. F. de la, … Valdimarsdóttir, U. A. (2018). Association of stress-related disorders with subsequent autoimmune disease. *JAMA*, *319*(23), 2388–2400. https://doi.org/10.1001/jama.2018.7028

Strum, Shirley. 2019. Why natural history is important to (primate) science: A baboon case study. *International Journal of Primatology*, *40*, 596–612.

Suchak, M., Eppley, T. M., Campbell, M. W., Feldman, R. A., Quarles, L. F., & Waal, F. B. M. de. (2016). How chimpanzees cooperate in a competitive world. *Proceedings of the National Academy of Sciences*, *113*(36), 10215–10220.

Tomasello, M. (2016). *A Natural History of Human Morality*. Cambridge, MA: Harvard University Press.

Tomasello, M., & Call, J. (2008). Assessing the validity of ape-human comparisons: A reply to Boesch (2007). *Journal of Comparative Psychology*, *122*(4), 449–452.

Tomasello, M., George, B. L., Kruger, A. C., Jeffrey, M., Farrar, & Evans, A. (1985). The development of gestural communication in young chimpanzees. *Journal of Human Evolution*, *14*(2), 175–186.

Tye, M. (2016). *Tense Bees and Shell-Shocked Crabs: Are Animals Conscious?* Oxford, UK: Oxford University Press.

Varner, G. E. (2012). *Personhood, Ethics, and Animal Cognition: Situating Animals in Hare's Two Level Utilitarianism*. Oxford, UK: Oxford University Press.

von Uexküll, J. (1957). A stroll through the worlds of animals and men: A picture book of invisible worlds. In *Instinctive Behavior: The Development of a Modern Concept*, edited and translated by Claire H. Schiller, New York: International Universities Press, pp. 5–80.

de Waal, F. B. M. (1999). Anthropomorphism and anthropodenial. *Philosophical Topics*, *27*(1), 255–280.

de Waal, F. B. M. (2003). Silent invasion: Imanishi's primatology and cultural bias in science. *Animal Cognition*, *6*, 293–299.

de Waal, F. B. M., Boesch, C., Horner, V., & Whiten, A. (2008). Comparing social skills of children and apes. *Science*, *319*(5863), 569–570.

van de Waal, E., Renevey, N., Favre, C. M., & Bshary, R. (2010). Selective attention to philopatric models causes directed social learning in wild vervet monkeys. *Proceedings of the Royal Society of London B*, *277*(1691), 2105–2111.

Walker, B. M. (2017). Jaak Panksepp: Pioneer of affective neuroscience. *Neuropsychopharmacology*, *42*, 2470.

Watson, D. (2019). The Rhetoric and Reality of Anthropomorphism in Artifical Intelligence. *Minds and Machines*, *29*(3), 417–440. https://doi.org/10.1007/s11023-019-09506-6

Weintraub, P. (2012). *Discover* Interview: Jaak Panksepp Pinned Down Humanity's 7 Primal Emotions. Retrieved April 3, 2020, from https://www.discovermagazine.com/mind/discover-interview-jaak-panksepp-pinned-down-humanitys-7-primal-emotions

Whiten, A. (1994). Grades of mindreading. In Lewis, C., & Mitchell, P. (Eds.), *Children's Early Understanding of Mind*, 277–292. Cambridge, MA: Cambridge University Press.

Whiten, A. (1996). When does smart behavior-reading become mind-reading? In Carruthers, P., & Smith, P. (Eds.), *Theories of Theories of Mind*, 277–292. Cambridge, MA: Cambridge University Press.

Whiten, A. (2013). Humans are not alone in computing how others see the world. *Animal Behaviour*, *86*(2), 213–221.

Whiten, A., & Byrne, R. (1988). Tactical deception in primates. *Behavioral and Brain Sciences*, *11*, 233–273.

Whiten, A., Goodall, J., McGrew, W. C., Nishida, T., Reynolds, V., Sugiyama, Y., . . . Boesch, C. (1999). Cultures in chimpanzees. *Nature*, *399*, 682–685.

Wynne, C. D. L. (2004). *Do Animals Think?* Princeton, NJ: Princeton University Press.

Wynne, C. D. L., & Udell, M. A. R. (2013). *Animal Cognition: Evolution, Behavior and Cognition (2nd Edition)*. New York, NY: Palgrave Macmillan.

Yoon, C. K. (2003, November 14). Donald R. Griffin, 88, dies; argued animals can think. *New York Times*. www.nytimes.com/2003/11/14/nyregion/donald-r-griffin-88-dies-argued-animals-can-think.html

Acknowledgments

This Element exists only because of the many scientists who have been willing to let a philosopher interloper into their labs, field sites, and conferences. In particular, I can't imagine this Element coming to light had it not been for the longtime support of Anne Russon and Sara Shettleworth. In 2006, York University awarded us a Seminar for Advanced Research grant, which led to the creation of the Greater Toronto Area Animal Cognition Discussion Group. This interdisciplinary gathering of philosophers and scientists is a testament to the value of interdisciplinary collaboration, and many projects have been born of it. I presented an early draft of this Element to the group and received valuable written feedback on the entire manuscript from Anne Russon, Sarah Shettleworth, Noam Miller, and Adam Bebko, and helpful discussions from Kevin Temple, Daphna Buchsbaum, Stefan Linquist, and my students Rebecca Ring, Brandon Tinklenberg, Dennis Papadopoulos and Tyler Delmore. Jonathan Birch also provided valuable comments, as did the audience at the Philosophy of Science Association meeting in Seattle. Finally, my deepest gratitude to Ronald Giere, who trained me to look at the ways in which science is practiced and how it is taught.

Cambridge Elements ☰

Philosophy of Biology

Grant Ramsey
KU Leuven

Grant Ramsey is a BOFZAP research professor at the Institute of Philosophy, KU Leuven, Belgium. His work centers on philosophical problems at the foundation of evolutionary biology. He has been awarded the Popper Prize twice for his work in this area. He also publishes in the philosophy of animal behavior, human nature and the moral emotions. He runs the Ramsey Lab (theramseylab.org), a highly collaborative research group focused on issues in the philosophy of the life sciences.

Michael Ruse
Florida State University

Michael Ruse is the Lucyle T. Werkmeister Professor of Philosophy and the Director of the Program in the History and Philosophy of Science at Florida State University. He is Professor Emeritus at the University of Guelph, in Ontario, Canada. He is a former Guggenheim fellow and Gifford lecturer. He is the author or editor of over sixty books, most recently *Darwinism as Religion: What Literature Tells Us about Evolution; On Purpose; The Problem of War: Darwinism, Christianity, and their Battle to Understand Human Conflict*; and *A Meaning to Life*.

About the Series

This Cambridge Elements series provides concise and structured introductions to all of the central topics in the philosophy of biology. Contributors to the series are cutting-edge researchers who offer balanced, comprehensive coverage of multiple perspectives, while also developing new ideas and arguments from a unique viewpoint.

Cambridge Elements ☰

Philosophy of Biology